INTERESTING

FACTS IN

HISTORY

Fun American Trivia

Check on page numbers

RANDALL STUEBER

Prepare to embark on a journey through the hidden corners of American history, where truth is often stranger than fiction. This book is your passport to discovering facts, some quirky, some bizarre, and some just very fascinating snippets of history that have shaped the United States. From the untold tales of Native American ingenuity to the peculiar events that never made it into your history textbooks, we've got over 1001 fun facts to share.

Each fact in this collection has been double-checked for accuracy, ensuring that you're getting the real deal. Whether you're a history buff, a trivia enthusiast, or someone who loves a good story, this book promises to entertain, enlighten, and maybe even surprise you.

Fun Fact: Did you know that during the Revolutionary War, the Oneida allies of George Washington's army walked hundreds of miles to deliver bushels of white corn to the starving soldiers at Valley Forge? Their knowledge and generosity helped save the army during a critical time.

So, buckle up and get ready to dive into America's lesser-known legends and truths. Who knows? You might just find yourself the star of your next trivia night!

Happy reading!

Interesting facts in history: Fun American Trivia

ISBN: 979-8-9870429-5-3 Paperback

CONTENTS

01

JUST FOR FUN

We have included some American Mythology for fun. Don't worry—you still have over 1,00o facts. .

1. Legend whispers of a man so mighty, his very footsteps carved out the Great Lakes. Paul Bunyan, the colossal lumberjack, alongside his loyal companion, Babe the Blue Ox, didn't just chop trees; they sculpted America's heartland with their legendary exploits. Could the contours of our land be the work of folklore come to life?

2. Step into the shoes of a barefoot mystic, a man whose legend grew with each sapling he planted. Johnny Appleseed, not just a tale, but a real-life nomad who traversed the American frontier, sowing seeds of apples and kindness, crafting a legacy that still blossoms today.

3. In the dense, whispering woods of North America lurks a shadow, a legend so enduring it's etched into the very soil— Bigfoot. In 1958, loggers stumbled upon footprints so colossal, they hinted at a creature beyond the known. Cryptic photos exist, yet Bigfoot remains elusive, a phantom of the forest, ever seen but never caught. What secrets does this giant of folklore guard within the trees?

4. Deep within New Jersey's Pine Barrens, where the shadows can trick the eye, a tale chills the spine. Since the 1700s, whispers of the Jersey Devil have haunted the night, a creature so bizarre it defies nature. A beast with the body of a kangaroo adorned with horns, tiny clawed hands, a forked tail, and eyes that glow like embers of hell. What ancient curse birthed such a monstrosity, and does it still prowl, waiting in the darkness?

5. In the early 19th century, in the remote woods of Tennessee, the Bell family encountered more than just rural solitude. From 1817 to 1821, they were besieged by an entity so malevolent, so real, it earned a name: The Bell Witch. This wasn't mere creaks in the night; it was a poltergeist that spoke, tormented, and left witnesses in utter dread. What dark force took residence with the Bells, and did it ever truly leave?

6. In the shadowed hills of West Virginia during the 1960s, sightings of a creature so otherworldly it could only be an omen gripped the town with fear. The Mothman, a being with wings and eyes that burned like red coals, was spotted just before calamity struck. Was this enigmatic entity a harbinger of doom, or a guardian with a cryptic message? The mystery lingers, as does the chilling question: What did its appearances foretell?

7. In 1897, Greenbrier County, West Virginia, witnessed a tale so eerie it crossed the very boundary between life and death. A spirit known as the Greenbrier Ghost returned from beyond, not in peace, but with a chilling mission. This was no ordinary haunting; this was a spectral quest for justice. The ghost of a murdered woman whispered the truth of her demise, leading to a conviction that still sends shivers down the spine. How did she reach from the grave to point the finger at her killer? The mystery of the Greenbrier Ghost remains one of the most haunting cases in paranormal lore.

8. Deep within Florida's murky swamps, where the air hangs heavy with mystery, lurks a creature so elusive, yet so potent in its presence, it's dubbed the Skunk Ape. Imagine Bigfoot, but with a twist—a scent so foul it's unforgettable. Sightings of this Floridian behemoth, with its pungent odor and colossal form, continue to perplex and thrill. As recent as January 2021, whispers of its haunting presence remind us that in the Everglades, something ancient and wild might still roam. What secrets does this swamp giant keep, and how close does it venture to our world?

9. Have you ever looked to the sky during a tempest and wondered about the origin of lightning? Native American legends speak of a creature so majestic and formidable, it commands the very elements. The Thunderbird, a colossal avian deity, is said to flap its mighty wings, summoning storms and wielding lightning like a weapon from the heavens. Is this ancient myth a mere tale, or does it whisper truths about the natural world we've yet to grasp? The Thunderbird watches from the clouds, a guardian or a harbinger, its story as electrifying as the storms it's believed to create.

10. Step into the shadowed corridors of the White House, where history meets the supernatural. Both Winston Churchill and Theodore Roosevelt, titans of the 20th century, reported encounters with a ghost so profound that it could only be that of Abraham Lincoln. Imagine the Prime Minister, or the Rough Rider President, coming face to face with the spectral figure of the Great Emancipator. Does the White House harbor more than just political secrets? Could spirits of past leaders linger, guiding or haunting those who follow? The ghostly tales of Lincoln suggest a White House where the past never truly rests.

11. In the heart of Lakota lore, where legends intertwine with the very fabric of existence, the emergence of a white buffalo is not merely an event, but a divine omen. This rare creature, its coat as pure as fresh snow, is believed to herald times of hope and renewal, a sacred sign from the spirits themselves. The white buffalo's appearance is more than a sighting; it's a prophecy, a beckoning towards a promising future.

12. John Henry is the legendary steel-driving man who was said to have hammered his way through a mountain faster than a steam-powered drill. According to folklore, he hammered so fiercely that he won the race against the machine but then collapsed and died from exhaustion. Talk about going out with a bang!

13. Then there is The Ghost of Bellamy Bridge: According to the lore, the ghost of Elizabeth Jane Bellamy haunts the historic

Bellamy Bridge in Marianna, Florida. Elizabeth was a young bride who heartbreakingly died on her wedding night in 1837. Elizabeth's spirit is said to be searching for her lost love.

14. A must see house. The Winchester Mystery House! This sprawling mansion in San Jose, California, was built by Sarah Winchester, the wealthy widow of William Wirt Winchester, who was a major player in the Winchester Repeating Arms Company. Legend has it that those killed by Winchester rifles haunted her. A medium told her she needed to build a home for these spirits. Sarah set out on a 38-year construction spree that resulted in a house with over 160 rooms, 47 fireplaces, and 10,000 panes of glass. Construction began in 1884.

15. Have you heard of Pecos Bill? He was said to have been raised by a pack of coyotes until a cowboy found him. Pecos Bill is created with creating the Rio Grande by dragging his lasso behind him. Once, he lassoed a tornado to stop it from destroying the prairie. He even shoots all the stars out of the sky, except the Lone Star of Texas.

16. In the shadowed tales of the Pacific Northwest, where myths weave into the air, there's a revered and feared creature— the Raven. Not just a bird but a deity in feathers, this trickster and creator is said to have stolen the light from its celestial  keepers, gifting it to humanity. Imagine a world shrouded in perpetual darkness until this cunning bird, with its gleaming eyes and sly wit, dared to challenge the gods. Did the Raven truly bring forth the dawn, or was it all part of an elaborate cosmic jest? The legends suggest a universe shaped by the whims of this enigmatic being, where every shadow might hide a story of creation or chaos.

17. Let's not pass the tail of Frogmen. The first sighting of the Loveland **Frogman** dates back to 1955 when a businessman claimed to see three frog-like creatures on the side of the road near the Little Miami River. Maybe they had stopped by the pub earlier in the day.

02

FOUNDING FATHERS

Some fun facts about the Founding Fathers.

18. Did you know that Morocco was the first country to recognize the United States as independent in 1777?

19. Many Americans don't know that New York City served as the capital of the United States from 1785 -1790.

20. What was the founding fathers' natural hair color? Many believe he wore a wig, but his hair was his own. He powdered it to achieve the fashionable white color at the time, concealing its true color—red! Yes, George Washington was a ginger!

21. Georg Washington was so worried about being buried alive that he insisted that they would wait at least three days before burying him.

22. What founding father grew pot? It was Washington! He grew marijuana at his Mount Vernon estate. Although it was used for industrial purposes at the time, it's still a fun fact to know.

23. Thomas Jefferson had a feathered friend who was none other than a mockingbird named Dick. The bird would often perch on Jefferson's shoulder while he worked.

24. Benjamin Franklin had – of all things - a pet squirrel named Mungo.

25. John Adams, the second President of the United States, decided to name his dog Satan. We do not know why.

26. Thomas Jefferson was fascinated by natural history and particularly interested in mastodons. Convinced that these prehistoric creatures still roamed the American wilderness, he instructed Lewis and Clark to look for them during their expedition.

27. Benjamin Franklin in the nude? As Franklin defined them, an air bath involved sitting naked with open windows, which he believed was good for his health. His neighbors most likely closed their windows.

28. Benjamin Franklin wasn't the only Founding Father who enjoyed "air baths." Thomas Jefferson also believed in the health benefits and would sit in his room completely naked to soak up the fresh air.

29. While it's well known that George Washington had dentures, what's less known is that his dentures were made from various materials, including human teeth, animal teeth, and ivory— NOT wood.

30. James Madison, the fourth President of the United States, was notably small in stature, standing at approximately 5 feet 4, nicknamed "Little Jimmy." Despite his physical stature, Madison played a crucial role in drafting the United States Constitution and establishing the Bill of Rights.

31. As a side note, the word "republic" appears in the United States Constitution, while "democracy" does not.

32. John Adams was an avid letter writer. He and his wife Abigail exchanged over 1,100 letters during their lifetime, and their correspondence is considered one of the greatest collections of letters in American history.

33. Benjamin Franklin showed he had a sense of humor when he wrote an essay titled "Fart Proudly" in 1781. In this humorous piece, Franklin suggested that scientists should work on making farts smell good.

34. Despite his wealth and status, Thomas Jefferson died deeply in debt. He was known for his lavish lifestyle and extravagant spending, which eventually caught up with him.

35. The Founding Fathers were known to enjoy their drinks. They often celebrated with alcohol, and it was common for them to be quite inebriated during important events.

36. Was there a president before George Washington? Well – sort of. Before the U.S. Constitution was ratified and George Washington became the first President of the United States under the new federal government, some individuals served as presidents of the Continental Congress and later the Congress of the Confederation. Two of these men were.

Peyton Randolph (1774-1775) was the first to serve as President of the Continental Congress.

John Hancock (1775-1777) is perhaps the most well-known due to his signature on the Declaration of Independence.

These individuals presided over what could be considered the first national government of the United States, though their role was quite different from the modern presidency.

37. Despite being political rivals, Thomas Jefferson and John Adams died on the same day, July 4, 1826, precisely 50 years after the Declaration of Independence was adopted. James Monroe also died on July 4th, though he died in 1831.

38. John Jay, one of the lesser-known Founding Fathers, was a big deal back then. He participated in negotiating the Treaty of Paris, which ended the Revolutionary War, and he was the first Chief Justice of the Supreme Court.

39. Rhode Island was the last state to approve the US Constitution, finally ratifying it on May 29, 1790.

40. Both the British and American sides used invisible ink to send secret messages. The ink could only be revealed by applying heat or a special chemical.

41. Do you know who was the only woman to earn a full military pension for her service in the Revolutionary War? It was Deborah Sampson. She disguised herself as a man and joined the Continental Army during the War. She served for over a year and a half before her identity was discovered.

42. When you think of Paul Revere, does the image of a patriot on horseback, galloping through the night, come to mind? But here's a twist to his tale that's as surprising as it is intriguing: Revere was also a man of the drill and the pliers—a dentist. Imagine the hands that warned of British invasion crafting dentures from the tusk of a walrus! This colonial multitasker didn't just alert the populace; he also alleviated their toothaches. How many smiles did he save, or perhaps, unwittingly altered with his ivory creations? Paul Revere: hero, artisan, and dental pioneer.

43. In the Revolutionary War, more soldiers died from disease than from actual combat. Smallpox, dysentery, and other illnesses were rampant in the camps.

44. Lemuel Cook served in the Revolutionary War at the age of 16 and lived to be 106. That's right, he was a teenager when the war started and was still kicking it when the Civil War broke out.

45. Many women served as spies during the war. One such woman was Anna Smith Strong! She was a key member of the Culper Spy Ring during the Revolutionary War. She used her laundry as a secret code system to communicate with other spies.

46. He is credited with saving George Washington twice. Hercules Mulligan was an Irish immigrant (and spy) who became a prominent tailor for the Redcoats in New York City. While measuring their inseams and waistlines, he would casually inquire about troop movements and plans. Twice, he found out about planes to assassinate George Washington and was able to pass on the information, saving Washington's life.

47. William Blount, one of the Founding Fathers and a signer of the Constitution was caught in 1797 conspiring to hand over parts of present-day Missouri and Louisiana to our pals across the pond. He planned to use Native Americans to do his dirty work. President John Adams got wind of the plot and promptly had Blount expelled from the Senate. It was the first time a senator had been kicked out of office.

03

US Presidents

───❖───

48. Before he donned the tall stovepipe hat of the 16th President, Abraham Lincoln wore many lesser-known crowns. Picture this: the future Great Emancipator, in the humble role of a postmaster in New Salem, Illinois, sorting letters with the same hands that would later pen the Emancipation Proclamation. However, his journey through professions was as varied as it was intriguing; Lincoln also wielded a surveyor's chain, measuring the land with a precision that would later map out his political path. And let's not forget his time as a lawyer, arguing cases in courtrooms, a practice ground for the debates that would shape a nation. Lincoln's early career was a tapestry of experiences that forged one of America's most revered figures.

49. Before he stitched together the threads of his political career, Andrew Johnson was quite literally a tailor in Greeneville, Tennessee. Imagine the hands that would one day sign executive orders, deftly sewing garments for his fellow townsfolk. What's more, Johnson never entirely left his craft behind. Amid debates and policy-making, he returned to his needle and thread, a president who could still hem a pair of trousers. This is the story of a man who tailored his destiny from the quiet of a tailor shop, proving that sometimes the humblest beginnings hold the fabric of greatness.

50. He graduated from the U.S. Naval Academy and served in the Navy before returning to Georgia to take over the family's peanut farming business before becoming **President** Jimmy Carter.

51. Has there been a president who served nonconsecutive terms? There have only been – 45 individual men to have served as President of the United States. And only one to have served non-consecutive terms. First, He served as Mayor of Buffalo, New York. Then, he was elected Governor of New York, where he vetoed bills he deemed wasteful or corrupt, making him popular but creating enemies within his Democratic Party. He became the 22nd and 24th President of the United States. He won the 1884 presidential election against James G. Blaine, becoming the first Democrat elected after the Civil War. He lost the 1888 election to Benjamin Harrison. In 1892, He defeated Harrison in a rematch, making Grover Cleveland the only president to serve two non-consecutive terms. Will Donald Trump become the 2nd person in history to have that distinction?

52. **First in Line, Last in Language**: President Martin van Buren, who served from 1837 to 1841, wasn't just the first president born a U.S. citizen; he also had a unique linguistic twist. Van Buren spoke Dutch as his first language, a nod to his Dutch-American heritage. Imagine the White House briefings with a touch of Dutch accent!

53. **From Silver Screen to Oval Office:** Before Ronald Reagan became the 40th President of the United States, he was a Hollywood heartthrob, starring in over 80 films. His role in politics began not with a script, but as the Governor of California, setting the stage for his presidency. Talk about a blockbuster career transition!

54. Have American presidents other than Trump been arrested in the past? One was arrested for running over an old woman. In 1853, while serving as the 14th President of the United States, Franklin Pierce was arrested for running over an old woman with his horse. Talk about a presidential faux pas! Despite the arrest, Pierce was not convicted. The courts at the time cited insufficient evidence, leading to the dismissal of any charges.

55. Do you know who worked as a mining engineer? Herbert Hoover became a millionaire through his engineering ventures before entering public service.

56. Harry S. Truman worked as a haberdasher, running a men's clothing store in Missouri before his political career took off.

57. Unelected but Not Uneventful: Gerald Ford, the 38th President, holds a unique spot in U.S. history as the only president who never won a national election for either the presidency or the vice presidency. Appointed to the vice presidency after Spiro Agnew resigned, He then ascended to the presidency following Nixon's resignation. A true political pinch-hitter!

58. From Classroom to Capitol: Before Lyndon B. Johnson became known for his political prowess, he shaped young minds as a teacher in Texas. Who knew that the future president was once correcting homework before crafting legislation?

59. The House's Unexpected Choice: In a twist worthy of political drama, 1824 marked the first time the House of Representatives had to pick the president. Andrew Jackson led the popular vote but lacked a majority. Under the Twelfth Amendment, the decision went to the House, which surprisingly elected John Quincy Adams. This controversial outcome fueled Jackson's fiery comeback, leading to his victory over Adams in the 1828 election. Talk about a political plot twist!

60. He reportedly owned over 80 pairs of pants. The 21st president of the United States (1881 to 1885): Chester A. Arthur was the James Bond of presidents when it came to fashion. Arthur believed that a well-dressed president was a sign of a well-run country.

61. The White House Rummage Sale: When President Chester A. Arthur moved into the White House, he decided it was time for a clean sweep. He organized a grand rummage sale, clearing out wagonloads of old furniture and decades of accumulated clutter. The sale netted him around $8,000, a hefty amount back in the day, turning the White House into a thrift store for a day!

62. Presidential Pop Culture: Before leading the free world, Barack Obama was a comic book enthusiast. As a kid, he collected issues of "Spider-Man" and "Conan the Barbarian," proving that even future presidents have a heroic side in their youth. Imagine him envisioning policies with the same fervor as Spider-Man swings through New York!

63. Sax Appeal on the Campaign Trail: Bill Clinton wasn't just smooth with policy; he had moves on the saxophone too. During his presidential campaign, he famously played the sax on "The Arsenio Hall Show," winning over voters with his musical talents. Who knew that the keys to the White House could also play a mean saxophone?

64. **Presidential Pins**: Richard Nixon had a unique way to unwind in the White House – he was keen on bowling. So much so, he installed a one-lane bowling alley and would often be found rolling strikes late into the night. Talk about a president with a penchant for pins!

65. **Truman's Tunes:** Harry S. Truman had a melody for stress relief; he was an accomplished pianist. Not content with just playing for relaxation, Truman once took to the stage for a concert in Washington, D.C., proving his presidency was not just about politics but also about hitting the right notes.

66. **Putting Politics Aside**: Woodrow Wilson was so passionate about golf that he couldn't resist bringing the game to the White House. He had a putting green installed on the lawn, ensuring he could try for a hole-in-one almost every day. Talk about combining governance with a good swing!

67. He played football in college and was offered contracts by two NFL teams. Gerald Ford was known for his athleticism and often swam in the White House pool.

68. Franklin D. Roosevelt collected stamps as a relaxing hobby. He had a vast collection and often spent time organizing and adding to it.

69. George Washington had a habit of dancing for hours at parties. He was known to be an excellent dancer and often led the dances at social gatherings.

70. James Buchanan, the 15th President, was the only bachelor president. His niece, Harriet Lane, was the White House hostess during his presidency.

71. Did he et stuck in the Tub? This new White House bathtub was 7 feet long, 41 inches wide, and weighed 400 pounds. His size led to some logistical challenges, including the need for a larger chair for his Supreme Court position after his presidency. He was the heaviest President, weighing around 330 pounds. He was rumored to have gotten stuck in the White House bathtub, so he installed an enormous bathtub to accommodate his size. We can't prove he got stuck, but he did order the enormous Bathtub. President William Howard Taft was the 27th president from March 1909 to March 1913.

72. It was over 100 years before he received his Medal of Honor. His courage was unquestionable, but some question whether his actions were extraordinary enough to warrant this honor. Others suggest that his criticism of the army secretary at the time might have influenced the decision not to award him the medal during his lifetime. So, was it withheld due to politics at the time? And why did Bill Clinton award him the Medal of Honor posthumously in 2001? We may never know for sure how much politics played in Theodore Roosevelt receiving the Medal of Honor 100 years after the fact. However, his bravery during the charge up San Juan Hill on July 1, 1898, was well-documented, and people were always significant in American history.

73. Despite his nickname "Silent Cal" for his quiet and reserved nature, Coolidge (the 30th President) had a playful side. He was known to press all the buttons on his desk to summon his staff, only to hide and watch their reactions.

74. Coolidge, an avid animal lover, had a variety of pets, including a raccoon named Rebecca, who walked on a leash around the White House grounds.

75. Coolidge even had a mechanical horse installed in the White House, which he would ride for exercise.

76. Zachary Taylor, the 12th President of the United States, was known for his impressive spitting skills, often aiming for a spittoon.

77. John Quincy Adams, the man who thought, "You know what the White House needs? A pet alligator!" So, the Marquis de Lafayette gifted him a scaly, toothy friend.

78. Politics get heated. James Monroe once chased a cabinet member out of the White House with hot fireplace tongs after a heated argument.

79. Lyndon B. Johnson often conducted meetings from the bathroom and even gave interviews while using the toilet—an interesting choice to multitask like that.

80. JFK was a speed reader! He could zip through a book at 1,200 words per minute, about four times the average reading speed!

81. Behind the composed facade of Rutherford B. Hayes, the 19th President of the United States, lurked a private terror known as lyssophobia—the fear of going mad. Imagine the man leading the nation, yet haunted by the specter of losing his own sanity. Hayes, who navigated the turbulent waters of post-Civil War America, carried this unseen burden, a silent companion to his every decision. How did this fear affect the policies he set, or the leadership he provided? The story of Hayes is not just one of political intrigue but of personal battle with the shadows of the mind.

82. Imagine being able to converse in a code that only you and your spouse understand, right in the open, yet utterly private. Herbert Hoover and his wife, Lou Henry Hoover, shared just such a secret. Fluent in Chinese, they would speak in this ancient tongue, their discussions veiled from prying ears. In the corridors of power, where every word can carry weight, this linguistic shield allowed them moments of intimacy and confidentiality. What whispers of policy, strategy, or perhaps just personal musings floated between them in a language few around could comprehend? The Hoovers' use of Chinese adds a layer of mystery to the 31st President's tenure, hinting at a life beyond the public eye, cloaked in the characters of an ancient script.

83. Barack Obama liked to brew beer in the White House. The beer, known as White House Honey Ale, was brewed using honey harvested from Michelle Obama's White House Garden.

84. Some historians have called her "the first woman to run the government." She had helped hide the president, who was evidently unable to fulfill his presidential duties for the last 17 months of his presidency. His wife, Edith, and his physician, Cary Grayson, maintained a veil of secrecy around his condition, keeping the public and the government in the dark. This led to conspiracy theories about who was running the country during this period. This is often cited as why he could not effectively manage the fallout from his ambitious foreign policy goals, which led to the U.S. NOT joining the League of Nations. We are talking about Woodrow Wilson, who, on October 2, 1919, had a stroke that left him mostly incapacitated.

85. Which president put the U.S. on the Gold Standard? President McKinley signed the Gold Standard Act in 1900, establishing gold as the only standard for redeeming paper money and stabilizing the economy.

86. President Grover Cleveland was the Sheriff of Erie County, New York, before becoming president. During his tenure, he executed two convicted murderers by hanging.

87. Before he navigated the corridors of power, Gerald Ford graced a different kind of stage—the modeling world. Imagine, the future 38th President of the United States not in the Oval Office, but on the cover of Cosmopolitan magazine. There he was, in April 1942, in the crisp lines of his Navy uniform, capturing the gaze of a nation in a different kind of spotlight. This brief foray into modeling might seem a light detour, but it adds an intriguing layer to Ford's persona—a man who, before leading the free world, was also an icon of style and patriotism. How did this early brush with fame shape the leader he would become? The story of Ford's modeling days is a curious chapter in presidential lore, blending glamour with governance.

88. We all know two men tried to assassinate Trump But do you know how many presidents have been Assassinated? Four sitting presidents have been assassinated: Abraham Lincoln (1865) by John Wilkes Booth, James A. Garfield (1881) by Charles J. Guiteau, William McKinley (1901) by Leon Czolgosz, and John F. Kennedy (1963) by Lee Harvey Oswald, (they claim?)

89. How many presidents have survived an assassination attempt? At least 7, if we're only counting the presidents who were physically harmed or nearly so.

Andrew Jackson was quite literally the first to say, "Not today," when someone tried to shoot him in 1835.

Theodore Roosevelt got shot in 1912 but decided to give a 90-minute speech anyway because, well, why not?

Franklin D. Roosevelt, before he was even president, during his campaign in 1933.

Harry S. Truman, in 1950, when two Puerto Rican nationalists tried to storm Blair House.

Gerald Ford, twice in 1975, by women.

Ronald Reagan in 1981, who got shot but quipped to his wife, "Honey, I forgot to duck." Classic Reagan.

And let's not forget Donald Trump in 2024, where the bullet grazed his ear. If that's not a close call, I don't know what is.

So, if we're counting both the successful and the "I'm still here" moments:

04

LANDMARKS &US MEMORIALS

---◆---

90. The Marble Majesty of Lincoln:

Step into the grand edifice of the Lincoln Memorial, where every block tells a story. While the serene visage of Abraham Lincoln, crafted from Georgia's pristine white marble, gazes solemnly over the Reflecting Pool, the structure itself is a marvel of material. The exterior, a testament to Colorado Yule marble, speaks of America's varied beauty. Yet, it's the statue within, weighing a monumental 175 tons, that anchors this shrine of democracy.

91. Washington's Time Capsule:

When laying the cornerstone of the Washington Monument in 1848, the builders buried more than just stones; they interred a piece of America's soul. Within this cornerstone lies a capsule, a secret archive containing relics like Washington's portrait, the Declaration of Independence, and even a Bible. Imagine, beneath the towering obelisk, a snapshot of a nation's infancy, waiting to be discovered.

92. The Lady Liberty's Hidden Face

The Statue of Liberty is not just an emblem of freedom but a canvas of personal artistry. The face, serene and watchful, was modeled after the sculptor's own mother, a poignant tribute. Yet, the body, strong and poised, reflects a different muse, blending familial homage with an idealized form.

93. Liberty's Secret Sanctuary:

High above the bustling island below, within Lady Liberty's torch, lies a chamber known to few. Once accessible, this secret room has been sealed since 1916 after an explosive incident. What mysteries or whispered secrets might this now-forbidden chamber hold, hidden from the world's gaze?

94. The Monument's Long March:

The Washington Monument's construction reads like a tale of perseverance. Begun in 1848, it stood incomplete for decades, its progress stalled by financial woes and the tumult of the Civil War. Only in 1884 did it reach its full height, a symbol of unity after division, a beacon of time that took nearly as long to build as it has stood complete.

95. The Secret Vault of Mount Rushmore:

Hidden behind the stoic visage of Abraham Lincoln on Mount Rushmore lies a chamber that few know of. This clandestine space contains a treasure trove not of gold, but of history— documents that preserve the essence of America's past. Imagine, nestled within the rock, a repository of our nation's soul, a time capsule for future generations to uncover the roots of their heritage.

96. The First Sentinel: Devil's Tower:

Rising from the plains of Wyoming, Devil's Tower is more than a geological wonder; it's a pioneer in preservation. Designated America's first national monument in 1906 by Theodore Roosevelt, this igneous marvel stands as a testament to the nation's commitment to conserving its natural treasures. Before parks were parks, Devil's Tower set the stage for the protection of America's scenic grandeur.

97. Miami's Mausoleum at Mary Brickell Park:

In the heart of Miami, Mary Brickell Park holds more than just verdant beauty; it cradles history. Here, the 1921 mausoleum of William and Mary Brickell is not merely a tomb but a monument to influence. This power duo, pivotal in shaping Miami's evolution, rest in a structure that whispers tales of the city's past, inviting visitors to reflect on the foundations of this tropical metropolis.

98. The Legacy of John Henry:

In Talcott, West Virginia, stands a bronze tribute to strength and spirit: the John Henry Monument. Unveiled in 1972 to celebrate the centennial of the Great Bend Tunnel's completion, this statue immortalizes the legendary steel-driver who, with hammer and heart, raced against the machine. For four decades, it marked the spot of his fabled contest until, in 2012, it was relocated to the John Henry Historical Park, where his story of human endurance against industrial might continues to inspire.

99. The Tri-State Rock:

Where New York, Pennsylvania, and New Jersey converge, lies a simple yet profound marker—the Old Tri-State Monument. This unassuming granite stone, known also as the Tri-State Rock, stands as a silent sentinel, delineating state boundaries with a touch of history. It's a quiet spot where three worlds meet, each state's jurisdiction ending where another begins, a testament to cartography and compromise.

100. The Newsboy Statue:

In Great Barrington, Massachusetts, a 19th-century tribute stands: The Newsboy Statue. Erected in 1895, this fixture honors the grit of the young newspaper vendors who once roamed the streets, their voices ringing through the air with the latest headlines. This bronze figure, with news in hand, serves not only as a reminder of a bygone era's labor but also as a salute to the resilience of youth in the face of daily hustle.

101. The Pigeon Hills Monument:

Nestled within Codorus State Park, Pennsylvania, the Pigeon Hills Monument stands as a poignant reminder of extinction. Erected by diligent Boy Scouts in 1947 to honor the passenger pigeon, this marker was later moved to its present location in the 1980s. It's not just a monument; it's a call to remember and reflect on the fragility of nature's wonders.

102. The Doris Miller Memorial:

In Waco, Texas, rises the Doris Miller Memorial, a tribute that echoes with courage. Dedicated in December 2017, on the anniversary of Pearl Harbor, this site commemorates Doris Miller, the first African American to receive the Navy Cross. His act of heroism during that fateful attack symbolizes the unyielding spirit of service against all odds.

103. The Hidden Wine Cellar of Brooklyn Bridge:

Beneath the iconic arches of the Brooklyn Bridge lies a secret not of steel and stone, but of spirits. During Prohibition, a clandestine wine cellar was concealed here, a quiet rebellion against the ban, storing vintages in the shadow of one of America's architectural marvels. It's a little-known tale of New York's spirited history, quite literally.

104. The Tiny Twin of the Washington Monument:

Hidden in plain sight on the grounds of the Washington Monument, beneath an unassuming manhole, lies a curious replica. This 12-foot-tall model, a miniature echo of its towering counterpart, adds an element of whimsy to the monument's history, inviting visitors to consider even the smallest details can hold grand significance.

105. The Lincoln Memorial's Secret Chamber:

Beneath the solemn gaze of Abraham Lincoln's statue lies a portal to another time—the Lincoln Memorial's hidden door. This inconspicuous entrance leads to a sprawling foundation space, its walls scrawled with the graffiti of the monument's builders, offering a time capsule of their thoughts and experiences.

106. The Burned at the Stake Monument:

In Wellesley, Massachusetts, stands an unusual tribute: the Burned at the Stake Monument. Dedicated to a college founder's supposed ancestor who faced martyrdom in 1555, this monument is a whisper from history, a reminder of faith's fiery trials, etched into the landscape of academia.

107. 'We Honor a Hero' Memorial:

In Tacoma, Washington, at Lowell Elementary School, the 'We Honor a Hero' Memorial stands as an eternal testament to Marvin Klegman. In 1949, this young boy's life was tragically cut short when he saved another during an earthquake. More than half a century later, in 2003, his bravery was commemorated, ensuring his act of heroism is never forgotten.

108. The National Mustard Museum:

Nestled in Middleton, Wisconsin, the National Mustard Museum is a quirky homage to mustard lovers everywhere. Boasting over 6,000 varieties from around the globe, this museum is more than a collection; it's a celebration of the zesty condiment with seasoned meals worldwide.

109. Vent Haven Museum:

For those intrigued by the art of ventriloquism, the Vent Haven Museum in Fort Mitchell, Kentucky, is a must-visit. Housing the world's largest collection of ventriloquist dummies, it's a silent but intriguing gallery where each figure seems to whisper tales of performances past. A place where you might keep your mouth closed but your ears open for the stories these wooden companions could tell.The Lincoln Memorial has **a hidden door** that leads to the building's foundation and contains a cavernous space with graffiti from the original construction workers.

110. The Mutter Museum:

Step into Philadelphia's Mutter Museum, where the bizarre meets the beautiful in medical history. Here, among such curiosities as slides of Albert Einstein's brain and a vast collection of 139 skulls, visitors embark on a journey through human anatomy's more unusual tales, a place where science meets spectacle.

111. The Thing in Dragoon:

In the heart of Arizona, The Thing in Dragoon has been a mystery since the '60s. This roadside enigma invites travelers to pay a small fee for a peek at the peculiar. From the mildly odd to the outright strange, it's a collection that promises to leave you with more questions than answers.

112. The Fremont Troll:

Beneath Seattle's Aurora Bridge lurks a local legend—the Fremont Troll. Since 1990, this monstrous sculpture has captured hearts, celebrating its birthday every Halloween with the community. It's more than just art; it's a quirky piece of Seattle's soul.

113. Seagull Monument:

In Salt Lake City, the Seagull Monument stands not just as a historical marker but as a tribute to nature's intervention. It commemorates the miraculous intervention of seagulls who saved early Mormon settlers from a cricket infestation, a tale of avian salvation.

114. Sh*t Fountain:

In Chicago, nestled in a not-so-fragrant corner, is the Sh*t Fountain—a bronze sculpture depicting a pile of dog excrement. Created by artist Jerzy S. Kenar, it's a pungent reminder for pet owners to clean up after their dogs, bringing humor to a common urban annoyance.

115. Pearl Harbor National Memorial:

Journey to Oahu, Hawaii, to the Pearl Harbor National Memorial, where history whispers through the waves. Overlooking the USS Arizona, this site is a poignant tribute to the lives lost in the 1941 attack, a symbol of peace straddling a sunken ship, inviting reflection on sacrifice and remembrance.

116. Castle Hill, Sitka:

In Alaska's Sitka, Castle Hill is more than a scenic overlook; it's where the stars and stripes first flew over Alaskan soil in 1867, marking America's acquisition from Russia. This site not only offers breathtaking views but also a window into a pivotal moment in American expansion.

THE CIVIL WAR ERA

117. We often recount the tales of Wild Bill Hickok, the legendary gunslinger of the Old West, but behind the iconic figure was a family steeped in courage of a different kind. William Alonzo Hickok, Wild Bill's father, was an abolitionist whose convictions transformed their Illinois home into a clandestine station on the Underground Railroad. In the shadow of his son's fame, this Hickok's legacy lies in the quiet act of aiding runaway slaves to freedom, making the family's home a beacon of hope in the fight against slavery.

118. The American Civil War, a nation torn apart, holds a somber record as the most lethal war in U.S. history. With estimates suggesting a death toll between 620,000 and 750,000 soldiers from both the Union and Confederate sides, this conflict not only reshaped the country but also left an indelible mark on American soil, its cost measured in lives forever lost.

119. The last living widow of a Civil War veteran, Helen Viola Jackson, did not pass away until December 16, 2020, at the age of 101. She married James Bolin, a Union soldier, in 1936 when she was 17 and he was 93.

120. The Union Army during the Civil War was a melting pot of America's diverse populace. At least one-third of its soldiers were immigrants, bringing a tapestry of cultures to the fight for union. Remarkably, nearly one in ten were African American, demonstrating not only the ethnic breadth of those who served but also the significant role Black soldiers played in the fight for freedom and unity.

121. Harriet Tubman was famous for her dedication to the Underground Railroad and for leading a raid during the Civil War that freed over 700 enslaved people.

122. Their youngest Civil War soldier was a child who was only 9 years old, while the oldest soldier was an 80-year-old man from Iowa.

123. At the Battle of Spotsylvania Court House in May 1864, a segment of the fighting known as the Bloody Angle witnessed one of the most ferocious engagements of the American Civil War. This brutal conflict reportedly claimed more American lives in a single day than were lost by American forces on Omaha Beach during the D-Day invasion in 1944. This comparison underscores the intensity and scale of the Civil War's bloodshed, highlighting a lesser-known but profoundly deadly chapter in American military history.

124. In an era when battlefields were considered men's domain, approximately 300 women defied convention by donning male attire to fight in the American Civil War. These brave individuals went to great lengths to serve their cause, disguising themselves as men to join the ranks, showcasing the lengths to which some would go for their beliefs and the desire to partake in the fight for their nation's future.

125. Most do not know this about the song "Dixie," as it is associated with the Confederacy, but it was written by a Northerner, Daniel Emmett, a loyal Unionist. Abraham Lincoln called it one of the best tunes he ever heard. Who knew?

126. Where did Memorial Day first come from? originally known as Decoration Day. Following the Civil War, numerous communities across the United States began decorating the graves of fallen soldiers. While many places claim the origin, Waterloo, New York, was officially recognized by President Lyndon B. Johnson in 1966 as the birthplace of Memorial Day due to its continuous observance since May 5, 1866, where the town closed businesses and decorated soldiers' graves.

127. The last verified Civil War veteran, Albert Henry Woolson, passed away on August 2, 1956. He served as a drummer boy in the Union Army and lived to the age of 106.

128. The last verified Confederate veteran, Pleasant Riggs Crump, died at the ripe old age of 104 in 1951. He served in the Confederate Army and witnessed the surrender at Appomattox.

129. In one of the Civil War's most daring escapades, Union soldiers under the leadership of James J. Andrews executed a bold heist in 1862. They stole the Confederate locomotive "The General" from Georgia in a bid to sabotage vital supply lines. What followed was a high-stakes chase, with Confederates hot on their heels aboard "The Texas." Although their mission ended in capture, the audacity of the Great Locomotive Chase remains a testament to the lengths to which both sides went to gain any advantage in the war.

130. In an innovative move to gain the upper hand during the Civil War, the Union Army in 1861 launched the Balloon Corps under the direction of Thaddeus Lowe. This pioneering unit used balloons for aerial reconnaissance, providing a bird's-eye view of Confederate positions. Lowe's aeronauts floated high above the battlefield, gathering intelligence that was invaluable for military strategy, marking one of the earliest uses of air power in warfare.

131. During the Siege of Petersburg in 1864, Union forces dug a tunnel under Confederate lines and filled it with explosives. The resulting explosion created a massive crater intended to break the Confederate defenses. However, the Union assault was poorly executed, leading to a disastrous defeat.

132. Confederate General Jubal Early's army was known for its ghostly appearances and disappearances in the Shenandoah Valley, earning them a reputation as a **phantom army.**

133. Colorado Gold Rush (1858-1861): Also known as the Pike's Peak Gold Rush, it began just before the Civil War and continued into the early years of the conflict.

134. In 1862, gold was discovered in Bannack, Montana, leading to a significant rush. The war didn't stop the miners.

135. The Battle of Hampton Roads in 1862 marked a pivotal moment in naval warfare, showcasing the first clash between ironclad warships: the Union's USS Monitor and the Confederate's CSS Virginia (formerly the USS Merrimack). This historic encounter ended without a clear victor, as neither ship managed to sink the other, but it revolutionized naval combat, highlighting the dawn of the ironclad era where wooden ships were obsolete against armored behemoths.

136. Battle of New Orleans (1862): Union forces under Admiral David Farragut captured New Orleans, giving the Union control of the Mississippi River.

137. Amid the chaos of the Civil War, Congress passed the Homestead Act in 1862, offering 160 acres of public land to settlers willing to live on and improve the property. This legislation is a beacon of opportunity. It was a chance to start anew, with the promise of landownership, even as the nation itself was divided.

138. Pacific Railway Act (1862): This legislation authorized the construction of the Transcontinental Railroad, bringing the eastern U.S. to the Pacific coast by train. This was a busy time in history.

139. Enacted in 1862, the Morrill Land-Grant Acts were groundbreaking in American education, providing states with federal lands to establish colleges specializing in agriculture and mechanical arts. This initiative not only transformed higher education by making it more accessible but also aimed to bolster the nation's agricultural and industrial capabilities, directly responding to the needs of a growing country amidst the Civil War tumult.

140. The International Committee of the Red Cross, founded in Switzerland, traces its origins to the harrowing scenes at the Battle of Solferino in 1859. Witnessing the immense suffering of wounded soldiers, Henry Dunant was inspired to create an organization dedicated to aiding those injured in war. Today, the Red Cross continues its legacy, offering humanitarian aid across the globe, a symbol of hope and relief in times of crisis.

141. In 1861, Richard Gatling invented the Gatling gun, one of the first successful rapid-fire weapons. However, it saw limited use during the Civil War.

142. The Trent Affair (1861): The Union Navy intercepted a British ship, the RMS Trent, and captured two Confederate diplomats. The British saw the seizure as a violation of their neutrality and an affront to their sovereignty. The confrontation almost started a war between the U.S. and Britain.

143. Creation of the idea of an Ambulance service. The Union Army established an organized **Ambulance Corps** in 1862, thanks to the efforts of Major Jonathan Letterman. He was the Medical Director of the Army of the Potomac.

144. Americana in São Paulo, Brazil, has a unique history due to its connection with American immigrants. After the American Civil War, many former Confederate sympathizers moved to *America, and their offspring still live there.*

145. **John Hay,** who is not well known (and should be), started his career as a private secretary to President Abraham Lincoln. He was very close to Lincoln and was present at his deathbed after the assassination. Hay served as the United States Secretary of State for President William McKinley as well as Theodore Roosevelt.

146. During the Civil War and before World War I, German was the most commonly spoken language in the United States after English.

147. Wilmer McLean had the Civil War knock on his front door...twice! First, the Battle of Bull Run in is front yard which started the Civil War. Then, when he thought he'd escaped the chaos by moving to Appomattox Court House, General Lee and General Grant, ready to wrap up the war, did it in his parlor.

148. The Libby Prison Escape is a testament to the indomitable spirit of the Union officers, making it one of the most significant prison breaks in American history! Richmond, Virginia, February 1864. Over 100 Union officers held captive in the infamous Libby Prison decided they'd had enough and were going to dig their way out—literally. On the night of February 9th, 1864, over 100 men made their daring escape. Some were recaptured, but many made it to the safety of Union lines.

149. The United States introduced its first federal income tax as a response to the financial demands of the Civil War, with the passage of the Revenue Act of 1861. This legislation imposed a 3% tax on incomes exceeding $800, a rate that might seem modest by today's standards but was groundbreaking at the time. Initially, this tax only affected a small portion of the population, as only about 3% earned above this threshold, equivalent to roughly $28,123 in today's dollars. However, this was not a permanent fixture; the tax was repealed post-war, only to be revisited with the ratification of the 16th Amendment in 1913. This amendment gave Congress the authority to levy an income tax without apportionment among the states, setting the stage for the modern income tax system we know today.

06

NATIVE AMERICAN

150. Squanto was a prominent member of the Patuxet tribe, he was captured by English explorers in 1614 but escaped to England, where he learned English and gained valuable knowledge about European culture that he would use to his advantage. Unfortunately, when Squanto returned to North America in 1619, he found that his tribe had been decimated by disease.

151. In 1621, **Squanto** helped the Pilgrims at Plymouth Colony survive by teaching them essential agricultural practices, such as planting corn and using **fish as fertilizer**. His guidance was instrumental in establishing a successful harvest, which led to the first Thanksgiving celebration.

152. Native Americans were not officially U.S. citizens until 1924, when the Indian Citizenship Act was passed, partly in recognition of their service in World War I.

153. The influence of the Algonquian languages: Many common English words, such as "caribou," "chipmunk," "moccasin," and "toboggan," come from the Algonquian languages. These languages are part of the larger Algonquian family, including over 20 languages, from Blackfoot to Cree.

154. Would Russia be interested in a stickball game Instead of war? Could we solve world conflicts this way? Indeed, Native Americans had wars between different tribes, but sometimes, they used different methods for Conflict Resolution. Instead of going to war, tribes sometimes resolved their conflicts through a game of Stickball. The outcome of the game could determine the resolution of disputes; winning or losing could influence political alliances, trade agreements, or social standings. It was a way to build or mend relationships between different groups without the bloodshed of actual combat. We should challenge Russia to stickball.

155. Known as the last wild Indian, Ishi lived most of his life outside modern culture but emerged from the wilderness in California in 1911 and became a living connection to the ways of the Yahi people, who were part of the larger Yana group.

156. Turtle Island: Many Native American tribes refer to North America as "Turtle Island," reflecting their creation stories and deep connection to the land.

157. The power of the written word. Sequoyah knew that the Cherokee culture was in danger, so he created a plan to help preserve his rich heritage. Sequoyah, born around 1770, recognized the power of written language to preserve Cherokee culture during a time of intense pressure from American expansion. He ingeniously created a syllabary with 85 characters, each representing a syllable in the Cherokee language. His groundbreaking work became a vital tool for cultural survival, helping the Cherokee maintain their identity and sovereignty amid efforts to assimilate them. Sequoyah's work not only preserved the Cherokee language but also inspired other Native American tribes to develop their own writing systems.

158. Was there a newspaper written in the Cherokee language? Remember Sequoyah, who created written language for the Cherokee people around 1809. He made it possible to have a newspaper written for the Cherokee people. It was established in 1828 and was the first Native American Newspaper. The newspaper was printed in both English and Cherokee, using the Cherokee syllabary developed by Sequoyah. And was named the "Cherokee Phoenix and Indians' Advocate.". Editors used the paper for news and as a platform for political discourse, advocating against the removal policies that would eventually lead to the Trail of Tears. It discussed political issues and covered Cherokee life's cultural, social, and economic aspects.

159. Who has the distinction of being the 1st native American to be appointed to a cabinet position? She is a member of the Laguna Pueblo tribe and hails from New Mexico. She served as a US Representative from New Mexico before her cabinet appointment And was appointed Secretary of the Interior in March 2021 by President Biden. This made Deb Haaland the first Native American to hold a cabinet position in U.S. history.

160. Originating from the Ojibwe (Chippewa) culture, dreamcatchers consist of a circular frame with a web-like pattern woven inside. Good dreams pass through the web, making their way to the person sleeping below, yet bad dreams are trapped and do not make it to the sleeping person.

161. **Potlatch ceremonies** are traditional gatherings practiced by Pacific Northwest Coast Indigenous peoples. These ceremonies serve as important social, economic, and political events where communities come together to celebrate significant occasions.

162. Born around 1768, Tecumseh was a fierce warrior and leader known for his powerful speeches that rallied his people. During the War of 1812, Tecumseh and his followers sided with the British, hoping that a British victory would help them reclaim their land. Unfortunately, the war didn't go as planned, and Tecumseh met his end at the Battle of Thames in 1813.

163. Crazy Horse, the Oglala Lakota leader known for his role in the Battle of the Little Bighorn, was initially named Cha-O-Ha, which means "Among the Trees" or "In the Wilderness." The name "Crazy Horse" was given to him later in life, reflecting his bravery and fierce fighting spirit.

164. Standing up to the government. Sitting Bull, a Hunkpapa Lakota leader, played a vital role in the Battle of the Little Bighorn. In 1883, during the opening of the Northern Pacific Railway, he gave a speech openly and bravely criticizing the treatment of Native Americans.

165. Sacagawea, the Shoshone woman who assisted the Lewis and Clark Expedition, was only 16 years old when she joined the expedition. She was a real go-getter, carrying her infant son, Jean Baptiste, throughout the journey.

166. Jim Thorpe is credited as the first Native American to win gold for the United States. Born in 1887, The Associated Press named him the most outstanding athlete of the first half of the 20th century. Thorpe won two gold medals in the 1912 Stockholm Olympics in the pentathlon and decathlon. His performance in these events was so dominant that King Gustav "V" of Sweden reportedly said to him, "You, sir, are the greatest athlete in the world."

167. Stand Watie was a Cherokee leader and Confederate general during the American Civil War and was the **only Native American** to be promoted to the rank of General in the Civil War.

168. Born in 1865, Susan La Flesche Picotte is credited as the first Native American woman to **earn a degree in medicine.** She provided healthcare to the Omaha people and opened the first non-government-funded reservation hospital.

169. Fred Begay, also known as Clever Fox, was a Navajo/Ute nuclear physicist and the first Native American **to earn a Ph.D. in physics.** In 1994, he received the NSF Lifetime Achievement Award.

170. Mary Golda Ross, born in 1908, was the first Native American female engineer and a force to be reckoned with in the aerospace industry. She worked on some seriously cool projects, like the P-38 Lightning, a fighter plane used during World War II. She was also part of the team that developed the Polaris missile system. It was designed as a submarine-launched ballistic missile (SLBM) to provide a nuclear deterrent that was less vulnerable to a first strike compared to land-based missiles or aircraft.

171. John Herrington, of the Chickasaw Nation, became the first Native American astronaut in 2002, flying on the Space Shuttle Endeavour.

172. Nicole Aunapu Mann, A member of the Wailacki of the Round Valley Indian Tribes, was a trailblazing astronaut who became the first Native American woman in space in 2022.

173. Charles Curtis, the first Native American to be Vice President of the United States, was born in 1860. He was a member of the Kaw Nation and had Osage ancestry. As a child, he lived on the Kaw Indian Reservation.

174. Deb Haaland, of the Laguna Pueblo tribe, made history as one of the first two Native American women elected to the U.S. House of Representatives in 2018.

175. In 2021, Deb Haaland became the first Native American to serve as a Cabinet secretary and lead the Department of the Interior.

176. Sharice Davids, of the Ho-Chunk Nation, became a member of the U.S. House of Representatives in 2018 alongside Deb Haaland.

177. Markwayne Mullin, of the Cherokee Nation, served in the House of Representatives from 2013 to 2023 and was elected to the U.S. Senate in 2022.

178. Mary Peltola: A Yup'ik woman, was elected to the U.S. House of Representatives in 2022, representing Alaska.

179. Who is Wilma Mankiller? She is the first woman Principal Chief of the Cherokee Nation, was inducted into the National Women's Hall of Fame (1993), and received the Presidential Medal of Freedom in 1988.

180. So, how many different Native American tribes are in the United States? As of January 2024, there are 574 federally recognized Indian tribes in the United States. This includes tribes from all over the country, with a significant number located in Alaska and California.

181. The largest Native American group in the US, by population, is the Navajo Nation, with 399,494 (and growing) enrolled tribal members as of 2021. They have surpassed the Cherokee Nation to become the largest recognized tribe in the country

182. Wovoka, a Northern Paiute spiritual leader born around 1856, sparked the Ghost Dance movement in the late 19th century. His vision promised to revive Native American traditions and bring back the buffalo, restoring the lands taken by settlers.

183. Maria Tallchief was born in Fairfax, Oklahoma, on January 24, 1925. She was a Native American ballerina of Osage Nation descent and is widely considered one of the greatest ballerinas of the 20th century. She was the first American to dance with the Paris Opera Ballet and was a prima ballerina with the New York City Ballet. She is remembered for her technical brilliance, her charismatic stage presence, and her commitment to promoting American ballet.

184. Chief Standing Bear of the Ponca tribe is remembered for successfully arguing in 1879 that Native Americans are "persons within the meaning of the law" and, therefore, have the right to habeas corpus.

07

PIONEERS AND EXPLORERS

185. During their famous expedition, Lewis and Clark captured a prairie dog and sent it to President Thomas Jefferson as a gift. The little critter survived the long journey and became a curiosity in Washington, D.C.

186. Daniel Boone, the legendary frontiersman, once rescued his daughter and two other girls after a group of Native Americans kidnaped them. Boone was later captured by the Shawnee in 1778 but managed to escape.

187. Boone founded Boonesborough, one of the first English-speaking settlements west of the Appalachian Mountains.

188. Davy Crockett was not just a frontiersman, but he was also elected to the US House of Representatives in 1827 and re-elected in 1833. One piece of Legislation he Proposed was to abolish the U.S. Military Academy at West Point, arguing that it used public funds to benefit the children of wealthy men.

189. The Alamo had nothing to do with the United States. It was Texas fighting for its independence.

190. Jedediah Smith, an explorer and fur trapper, led a group through the South Pass west of the Rocky Mountains in 1824, essentially blazing the Oregon Trail. He was one of the first Americans to cross the Mojave Desert.

191. Jedediah Smith was crazy tough. Smith was attacked by a grizzly bear in 1823, which nearly tore off his scalp and ear. He had his companions sew them back on, and he continued his explorations.

192. In the 1830s, James Beckwourth, an African American mountain man, was adopted by the Crow Nation and became a chief after helping in a battle against the Blackfeet tribe. He discovered Beckwourth Pass, a key route through the Sierra Nevada.

193. The first person to map the South Pass through the Rocky Mountains was John Frémont In 1842.

194. Kit Carson: In the 1840s, Carson was a mountain man, scout, and guide. He was illiterate for most of his life but learned to read and write later in his career.

195. One of the longest wagon trains took part in the Great Emigration of 1843 on the Oregon Trail. This massive wagon train included over 120 wagons and around 1,000 people.

196. In the treeless plains, pioneers often used dried buffalo dung, known as "buffalo chips," as fuel for their fires. It must have given the camp a less-than-favorable fragrance.

197. Pioneer women used perfume made from natural ingredients like flowers, herbs, and spices to mask body odors, as regular bathing was not always possible on the frontier.

198. The average covered wagon measured around 10 feet long and 4 feet wide and was generally 7-8 feet tall.

199. During the Oregon era, a covered wagon cost around $85. The canvas cover could add more than $100. Outfitting, it was more.

200. Many emigrants were driven by what was termed the **"Oregon Fever,"** a kind of mass hysteria fueled by glowing reports of the fertile lands in Oregon.

201. At night, the pioneers would form their wagons into a circle to create a makeshift stockade. This was primarily to protect against potential Native American raids, although such attacks were relatively rare.

202. Settlers used the Oregon Trail from 1839 to 1869. During that time, an estimated 20,000 to 30,000 people died while traveling to Oregon, with about 20 graves per mile.

203. Narcissa Whitman was one of the first white women to travel on the Oregon Trail in 1836.

204. At 70 years old, Sarah Keyes was one of the oldest members of the 1843 migration.

205. So here's an interesting fact: at the same time, people were struggling on the Oregon trail in covered wagons. Scottish inventor Alexander Bain received a patent for what is often considered the earliest fax machine.

206. The most infamous story of pioneer hardship is that of the **Donner Party.** They faced starvation because they were trapped by snow in the mountains of Sierra Nevada during the winter of 1846-1847. some members resorted to cannibalism to survive.

207. Pioneers had to be resourceful with their food. They often had "hardtack," a type of hard, dry biscuit that could last for years. It was so tough that some joked it could double as a weapon or building material.

208. After his time on the Lewis and Clark Expedition, Colter became one of the original mountain men of the American West.

209. **Colter's Run**: In 1809, Colter was captured by Blackfeet warriors. They stripped him naked and gave him a short head start before chasing him like wild game. Colter managed to outrun most of the warriors, disarm and kill his closest pursuer, and then trek over 200 miles to safety, clothed only in a blanket.

08

WILD WEST

210. It was in 1836, that Samuel Colt invented the Colt Paterson revolver, the first usable six-shooter in the United States.

211. Wild Bill Hickok's preferred firearm was the Colt 1851 Navy revolver. He often carried a pair of these .36-caliber, six-shot revolvers, which were typically ivory-handled and added a distinctive touch to his iconic image.

212. The Colt Single Action Army Revolver (Peacemaker) was one of the most iconic guns of the Old West, favored by many lawmen, including Wyatt Earp and Bat Masterson.

213. The first large-scale cattle drive in the Old West occurred in 1866. Several Texas ranchers rode together to drive approximately 200,000 head of cattle north to the nearest railhead at Sedalia, Missouri.

214. The Chisholm Trail was one of the most famous cattle-driving routes in the Old West. It was named after Jesse Chisholm, a half-Cherokee trader who never actually drove cattle on the trail himself.

215. Cowboys on cattle drives often ate a diet consisting mainly of beans, biscuits, dried meat, and coffee. Fresh meat was a rare treat, usually only available when an animal was injured or died.

216. The Pony Express delivered its first mail on April 3, 1860, when it left St. Joseph, Missouri. Riders had to take a loyalty oath promising not to curse, drink, or fight.

217. The youngest Pony Express rider was "Bronco" Charlie Miller, who claimed he was only 11 years old.

218. Rutherford B. Hayes, the first sitting president to visit the West Coast of the United States, made his historic trip in 1880, visiting cities such as Seattle and San Francisco.

219. *Why did the cowboy ride his horse into town? Because it was too heavy to carry!*

220. In public places like saloons, using a communal towel to wipe one's face or hands was common. Yuk!!

221. The whiskey served in many early saloons was a potent mix of raw alcohol, burnt sugar, and chewing tobacco. It was often referred to by colorful names like "coffin varnish," "tanglefoot," and "rotgut."

222. The largest nugget of silver ever mined weighed a whopping 1,840 pounds! It was discovered in the Smuggler Mine near Aspen, Colorado, in 1894.

223. You know those iconic swinging doors, known as batwing doors? They were not just for show. They allowed for ventilation and easy access while maintaining some privacy.

224. The Bird Cage Theatre In Tombstone, Arizona, was both a saloon and a theater. It was known for its wild entertainment and was open 24 hours a day, seven days a week.

225. A shot of whiskey typically costs around 10 to 25 cents. The quality of the whiskey could vary greatly.

226. A glass of beer usually costs about 10 cents. Due to its short shelf life and high transportation costs, many saloons brewed their own beer, similar to microbrews today.

227. While primarily known as a cattle rancher and trailblazer (the Goodnight-Loving Trail in 1866), Charles Goodnight also served as a judge in the Texas Panhandle.

228. Known as the "Law West of the Pecos," Judge Roy Bean was a saloon keeper turned justice of the peace in Pecos County, Texas. Despite having no formal legal training, he became famous for his unconventional and often humorous rulings, which he delivered from his saloon. In one of his most famous rulings, Bean fined a dead man $40 for carrying a concealed weapon.

229. Wells Fargo began its banking operations in the West on March 18, 1852. Founded by Henry Wells and William Fargo, the company was established to provide banking and express services during the California Gold Rush. So they have been around a long time.

230. The primary purpose of many stagecoach routes was to transport mail and money. Passengers were just an additional source of revenue.

231. Why did the cowboy buy a dachshund? To get a long, little doggie!

232. The term "riding shotgun" originated from the practice in the Old West where an armed guard, often carrying a shotgun, would sit next to the stagecoach driver to protect against bandits and other threats. This guard was known as a "shotgun messenger."

233. The Chicken Bandit: A man named "Chicken Bill" Lovell tried to steal a chicken by hiding it under his coat. Unfortunately for him, the chicken started squawking, giving him away.

234. In the Old West, some settlers believed that placing a silver coin in their water barrels would help keep the water fresh and prevent bacteria growth.

235. Many settlers used herbs like willow bark (for pain relief), peppermint (for digestive issues), and echinacea (for infections).

236. Castor Oil: This was a popular remedy for constipation and other digestive issues. It was often included in settlers' medical supplies.

237. Poultices and Plasters: Made from ingredients like mustard, onions, or bread, they were applied to the skin to draw out infections or reduce inflammation.

238. Buffalo Bill Cody once staged a mock battle between cowboys and Native Americans in England for Queen Victoria's Golden Jubilee in 1887. The Queen was reportedly so impressed that she requested multiple encores.

239. The iconic cowboy hat, popularized by Hollywood, wasn't as common as you might think. Many cowboys wore bowler hats or derby hats because they were less likely to blow off in the wind.

240. The famous outlaw Billy the Kid was long thought to be left-handed because of a well-known tintype photograph showing him with a gun belt on his left side. However, it turns out that most tintype cameras produced a reversed image, meaning the photo was a mirror image. This means Billy the Kid was right-handed.

241. Where did the name "Death Valley" come from? In 1849, a group of pioneers known as the Lost '49ers attempted to take a shortcut through Death Valley on their way to California. They became hopelessly lost, and many perished in the harsh conditions. The survivors eventually made it out, and as they left, one reportedly said, "Goodbye, Death Valley," giving the valley its ominous name.

242. Where did the name "Tombstone" come from? In 1877, a silver prospector named Ed Schieffelin set out from Camp Huachuca in southeastern Arizona to explore the Dragoon Mountains. The soldiers at the camp warned him that the only thing he would find 'out there' was his own tombstone. Defying their warnings, Schieffelin struck silver and decided to name his mine "Tombstone" in a nod to the soldiers' grim prediction. The town that sprang up around his successful mine adopted the same name.

243. Another quirky tale involves the U.S. Camel Corps. In the mid-1800s, the U.S. government imported camels to help with transportation in the arid Southwest. These camels, however, were not well-received and eventually roamed wild.

244. The legend of the Red Ghost—a camel that roamed the Arizona Territory with a skeleton strapped to its back. It was a source of fear and fascination for many settlers but was eventually shot and killed. When the camel's body was examined, the skeletal remains of a human were found still strapped to its back. Some legends turn out to be true.

245. Unlike what most people think, many forts in the Wild West were not fully stockaded, with walls all around them.

246. There were two newspapers in Tombstone with opposing political views. "The Tombstone Epitaph," The Republican newspaper, And "The Nugget," The Democrats newspaper.

247. In 1874, Joseph Glidden invented barbed wire, which transformed agriculture by providing an effective means of fencing large land areas. This created some fighting in the West.

09

LAWMEN/WOMEN

248. A personal favorite – Wyatt Earp – mostly known for the Gunfight at the OK Corral that took place on October 26, 1881. Three cowboys were killed, but somehow Wyatt avoided being shot.

249. Ike Clanton ran away from the Gunfight at the OK Corral but was killed a couple of years later by another lawman while attempting to flee arrest for cattle rustling. Old habits are hard to Break.

250. Wyatt Earp owned saloons but did not drink as he liked sarsaparilla (root beer). Maybe that's why he fared so well in all those gunfights.

251. Wyatt lived to be 80 years old and passed on January 13, 1929. In all those years and gun fights, he had never been shot, so he lived a long life.

252. Wild Bill Hickok is most famous for the Dead man's hand. The poker hand he was holding when he was shot in the back and killed in 1865. He was holding Ace's and eights.

253. The first known quick-draw gunfight in the United States occurred on July 21, 1865, in Springfield, Missouri. It was a showdown between "Wild Bill" Hickok and Davis Tutt over a gambling debt and a prized watch.

254. Bat Masterson, a legendary figure of the Wild West known for his exploits as a buffalo hunter, army scout, gambler, lawman, and gunfighter, did not wear the customary cowboy hat but is often depicted wearing a derby hat and carrying a cane, giving him a dapper appearance.

255. Allan Pinkerton founded the Pinkerton National Detective Agency in 1850. Moto was "We Never Sleep"

256. Plot to kill Lincoln before he was President: In early 1861, Abraham Lincoln was preparing to travel from Springfield, Illinois, to Washington, D.C., for his inauguration, and there was no Secret Service at this time in history. Allan Pinkerton is known for starting the Pinkerton Detective Agency. He uncovered a plot to assassinate Lincoln in Baltimore, Maryland, as he changed trains there. Pinkerton and key figures devised a secret plan to alter Lincoln's travel schedule. They arranged for Lincoln to pass through Baltimore incognito and earlier than announced. Bypassing the planned route and public appearances, he arrived safely in Washington, D.C., thus thwarting the assassination attempt.

257. Kate Warne was the first female Pinkerton detective hired by the National Detective Agency in 1856.

258. The Pinkertons still exist today and are involved in Cybersecurity, amongst other things.

259. Famous for the wrong reason. Henry Plummer served as the sheriff of Bannack, Montana, in the early 1860s, but he was also the head of a gang of outlaws known as the "innocents" who were responsible for numerous robberies and murders. In 1864 a group of Vigilantes hung him without the benefit of a trial. They didn't like their sheriff.

260. The "Hanging Judge" of the Old West: Judge Isaac Parker was a federal judge in Arkansas from 1875 to 1896. During his tenure, he sentenced 160 people to death, of - 79 were executed.

261. The "Three Guardsmen": U.S. Marshals Heck Thomas, Chris Madsen, and Bill Tilghman were lawmen who took down several outlaws, including the Doolin Gang (the wild bunch).

262. Only a few women were in law enforcement in the Wild West. One of the few in 1893 was Ada Curnutt, a deputy U.S. Marshal in Oklahoma.

263. In the 1920s, Catherine "Cattle Kate" Jones became a deputy sheriff in Cave Creek, Arizona. She was a small woman who packed a pistol and fought bootleggers during Prohibition.

264. Another notable figure was Bass Reeves, born into slavery, he escaped to the Indian Territory and was later appointed deputy Marshal, becoming one of the first black deputy U.S. marshals west of the Mississippi River. He is credited with arresting over 3,000 outlaws and killing 14 in self-defense during his career.

265. Perhaps the most famous lawman of the 1920s and 1930s was Eliot Ness, a Prohibition agent known for his goal of bringing down Al Capone in Chicago. He led a team of agents known as "The Untouchables." Despite the danger of his job, he seldom carried a gun.

266. From the Gunfight at the OK Corral to being on TV in one generation. Wyatt Earp had no children, but his older brother Virgil Earp had a son. Virgil Earp Jr became a Lawman in the 1940s following the family legacy. In 1958, Virgil Earp Jr. appeared on the popular quiz show "The $64,000 Question," where he answered questions about the Wild West.

267. In 1910, Alice Stebbins Wells was the first female police officer who had the power to make arrests in the United States, serving in the Los Angeles Police Department.

268. 1910: The Mann Act, which prohibited interstate transportation of women for immoral purposes, was one of the first major laws enforced by the Bureau of Investigation (BOI) (precursor to the FBI).

269. 1924: J. Edgar Hoover was appointed as the director of the BOI. He would go on to lead the agency for nearly 48 years, significantly shaping its development.

270. In 1919, Georgia Ann Robinson became the first African American female police officer in the Los Angeles Police Department. Her career as a police officer ended in 1928 after a serious head injury she received while trying to break up a fight between two inmates.

271. In 1912, Isabella Goodwin became New York City's first female detective. She was known for her undercover work.

272. Virginia Hall, the first female CIA agent, joined in 1947, shortly after its formation. Despite losing part of her leg in a hunting accident, she continued her espionage work with a prosthetic leg. The Germans referred to her as "The Limping Lady" and considered her one of the most dangerous Allied spies.

273. A New York City police officer (Frank Serpico) exposed corruption within the NYPD. This led to the formation of the Knapp Commission, which investigated police corruption. His story was depicted in the 1973 film Serpico, starring Al Pacino.

274. Daryl Gates, often referred to as the "Father of SWAT," was the Chief of Police (LAPD)from 1978 to 1992, and during his tenure, he initiated the creation of the Special Weapons and Tactics (SWAT) team.

275. Alaska P. Davidson was the first female FBI agent. She was appointed as a special agent by then-director William J. Burns on October 11, 1922.

276. Alongside Susan Roley Malone, Misko was one of the first two women to graduate from the FBI Academy in 1972. She was formerly a nun.

277. In 1976, Sylvia Mathis became the first African American woman to serve as an FBI special agent.

278. The first female special agents in the U.S. Secret Service were Laurie Anderson, Sue Ann Baker, Kathryn Clark, Holly Hufschmidt, and Phyllis Shantz. They were sworn in on December 15, 1971.

279. The FBI employs over 10,000 special agents and over 20,000 professional staff members.

ODD LAWS AND REGULATIONS

280. "The Committee to End Pay Toilets In America (CEPTIA) was founded in 1970 by four high school students in Chicago. They campaigned on the basis that pay toilets were unfair to women (since men could use urinals for free), and their activism led to bans in many cities. By the mid-1970s, several states and municipalities had banned pay toilets.

281. Since 1869, New Jersey law has required bicycles to have bells that can be heard from at least 100 feet away.

282. In New Jersey, a 19th-century law allowed homeowners to kill a neighbor's pig if it wandered onto their property, provided the carcass was sent to a county overseer.

283. In Arizona, do not let your donkey sleep in the bathtub. It's illegal.

284. In Pennsylvania, using dynamite to catch fish is illegal.

285. No Ice Cream Cones in Pockets (Kentucky): carrying an ice cream cone in your pocket is illegal. This law was meant to prevent horse theft, as people would use ice cream to lure horses away.

286. Wearing a fake mustache, which causes laughter in church In Alabama, is illegal.

287. How is this enforced? In Vermont, Whistling underwater is prohibited.

288. What a blast. In Massachusetts, selling exploding golf balls is illegal. You can be fined up to $500 for the first offense.

289. In Kennesaw, Georgia, according to a city ordinance passed in 1982, If you are the head of the household in the city limits, you are required to maintain a firearm.

290. In New York, if a house is reputed to be haunted, the seller must disclose this to potential buyers.

291. Do not disrobe if a portrait of a man is nearby. In Ohio, it is a third-degree misdemeanor.

292. In Galesburg, Illinois, it's illegal to keep a smelly dog.

293. A woman's husband is required to walk in front of the car while waving a flag as she drives it. This was a law in both Louisiana and Virginia.

294. In International Falls, Minnesota, it's illegal for cats to chase dogs up telephone poles.

295. In Arizona, it is illegal to drive a car in reverse on a public road.

296. Let's be quiet; people are eating. In Little Rock, Arkansas, honking your car horn close to places that serve cold drinks or sandwiches after 9 p.m. is against the law.

297. In Glendale, California, jumping from a car going 65 mph or faster is illegal. Does that mean that if the car is driving at 64 mph, it is legal?

298. In Carmel, California, it's illegal to wear high heels over two inches in height without a permit.

299. In Missouri, bear wrestling is explicitly illegal. Violating this law is considered a class A misdemeanor.

300. In Alabama, it's illegal to play dominoes on Sundays.

301. In 1839, New York passed a law disqualifying duelists from holding public office. If you're going to kill each other, you can't run for office.

302. In the early 1900s, several cities, including Chicago, passed laws limiting hatpins to 9 inches for safety reasons.

303. In Kansas, it was illegal to eat ice cream on cherry pie in the 1800s. I could not find any cases of it being enforced.

304. In Vermont, a woman needed her husband's **written** permission to wear false teeth.

305. In Kansas, it was illegal to serve cherry pie à la mode on Sundays.

306. It's illegal to hunt camels in Arizona.

307. In Indiana, it's illegal for a man to have a mustache if he tends to kiss other humans habitually.

308. No throwing a ball at someone's head for fun: This law was put in place to protect carnival workers, though the exact year is unclear.

309. No slippers after 10:00 PM: New York City has a law that prohibits wearing slippers in public establishments after 10 PM.

310. No selfies with tigers: This modern safety law was enacted in 2014.

311. In Los Angeles, it's illegal to play frisbee on the beach without the presence of a lifeguard.

312. New York has a quirky law prohibiting citizens from greeting each other by "putting one's thumb to the nose and wiggling the fingers." This gesture, known as "cocking a snook," is considered a sign of derision and was deemed offensive enough to warrant a law against it.

313. In San Francisco, California, using used underwear to wipe your car is illegal.

314. In the great state of Alabama, there's a quirky law that prohibits the opening of an umbrella on the street. It's about keeping the horses calm and collected.

315. Oklahoma—You must tether your car outside of public buildings. There is no guidance on what to tether it to.

316. Washington—A motorist with criminal intentions must stop at the city limits and make a call to the chief of police as he enters the town.

317. Florida – It is illegal to skateboard unless you have a license.

318. In Florida, there's a quirky law that makes it illegal to sleep under a hair dryer at a hair salon. Both the woman and the salon owner can be fined for this offense.

319. Florida is known for its unusual laws. For example, did you know in public places it's illegal to sing if you are wearing a swimsuit?

320. Illinois—It's a law that your car must have a steering wheel, or you can't drive it.

321. South Carolina—When approaching a four-way or blind intersection in a non-horse-driven vehicle, you must stop 100 ft. from the intersection and discharge a firearm into the air to warn horse traffic.

322. South Dakota—Horses must wear pants to enter a town named the Fountain Inn.

323. This was a law in Florida until 2005. Unmarried women are not allowed to parachute on Sundays.

324. In Minnesota, Crossing state lines with a duck on your head is illegal.

325. In Montana, It's illegal to give a rat as a present. But you can give a rat as food?

326. In Galesburg, Illinois, engaging in "fancy riding" on a bicycle is illegal. This means you can't remove both hands from the handlebars, take your feet off the pedals, or perform any acrobatic tricks while riding on city streets.

327. In Northfield, Connecticut, it was illegal to eat raw onions while walking down the street.

328. In Nacogdoches, Texas, young women were not allowed to have raw onions after 6 p.m.

329. In Oklahoma and Ohio, it was illegal to make faces at a dog.

330. In California, it was illegal to eat a frog that died during a frog-jumping competition.

11

LEGAL STUFF?

331. The first Supreme Court of the United States, established by the Judiciary Act of 1789, consisted of six justices. They were appointed by President George Washington and confirmed by the Senate.

332. The Court held its first session on February 1, 1790.

333. Robert Harrison was too ill to attend the first session and resigned shortly after his appointment.

334. William Cushing was the only justice to wear a white wig to court, a tradition he carried over from his time on the Massachusetts bench.

335. Nine justices were on the U.S. Supreme Court in 1869. Before that, the number fluctuated several times for political and practical reasons.

336. The Judiciary Act of 1869, signed into law by President Ulysses S. Grant, established the number of justices at nine, where it has remained ever since.

337. President George Washington appointed the first U.S. Marshals on September 26, 1789, following the passage of the Judiciary Act of 1789.

338. At one time, women could not sue or be sued in America. This started to change in 1839 with the passage of the "Married Women's Property Acts," Starting in Mississippi. This allowed women to own land and to sue, as well as to sue in their 'own' name.

339. Has politics always been a blood sport? Yes, yes, it has. Let's talk about two notable politicians from the past. The first was Thomas Hart Benton. Before becoming a senator, he had two duels, despite them being illegal, with a fellow Attorney and political rival, Charles Lucas. He killed Luca in the second duel. The duel did not significantly derail Benton's political career. He continued to serve as a Missouri Senator from 1821 to 1851, a true testament to his endurance in the political.

340. A second politician was Andrew Jackson. He also killed a man (Charles Dickinson) in 1806 in an illegal duel. Again, the duel did not harm Jackson's political career. In fact, it may have enhanced his reputation as a man of honor and courage, traits that were highly valued in frontier society of the time. He did, after all become President of the United States on March 4, 1829, So yes, politics has long been a blood sport.

341. The most famous American duel was where the Federalist and former secretary of the treasury Alexander Hamilton was killed in a duel with Aaron Burr, the vice president under Thomas Jefferson. Politics were rough back then.

342. Female marshals in the US are not new. You may believe it was not until recently that women were involved in law enforcement. However, the first female federal marshal in the US was Phoebe Couzins, who also accomplished many things. She was appointed a Deputy U.S. Marshal for the Eastern District of Missouri in 1887. She is also credited as one of the first female lawyers in the United States.

343. Belle Boyd: A Confederate spy during the American Civil War used her charm to gather intelligence from Union soldiers. She began her spying activities at the age of 17 after witnessing a Union soldier strike her mother. She shot and killed the soldier, which led to her first arrest; though she was later released, Belle was arrested multiple times but always managed to charm her way out or escape.

344. The United States Secret Service was established on July 5, 1865. Its purpose was to combat widespread counterfeiting, but the Secret Service was not yet responsible for protecting the president.

345. The Secret Service began protecting the President of the United States following the assassination of President William McKinley in 1901.

346. The FBI was founded on July 26, 1908. It began as a small group of special agents within the Department of Justice, initially called the Bureau of Investigation. It was later renamed the Federal Bureau of Investigation (FBI) in 1935.

347. United States v. Forty Barrels and Twenty Kegs of Coca-Cola (1916): This case involved the government suing Coca-Cola over the caffeine content in its drinks.

348. Sandra Day O'Connor became the first female Supreme Court justice in the United States in 1991. She was an appointee of President Ronald Reagan and served from 1981 to 2006.

349. An odd FBI investigation was The Toynbee Tiles. These mysterious tiles, embedded in asphalt in various cities, contain cryptic messages about resurrecting the dead-on Jupiter. The FBI investigated but never solved the mystery of who was behind them.

350. Operation Midnight Climax: In the 1950s and 60s, the CIA, with some cooperation from the FBI, ran a series of experiments where they lured unsuspecting individuals to safe houses and dosed them with LSD to study mind control.

351. The Isabella Stewart Gardner Museum Heist: In 1990, two men impersonation police officers got away with 13 pieces of art worth over $500 million from the museum in Boston. Despite extensive FBI investigations, the artwork has never been recovered.

352. The CIA was created under the National Security Act of 1947, which President Truman signed into law on July 26, 1947. The idea behind creating the CIA was to avoid another attack like Pearl Harbor in 1941.

353. Burnita Shelton Matthews was appointed to the Federal District Court for the District of Columbia as the first female federal judge in the United States in 1949, by President Harry S. Truman.

354. Julius and Ethel Rosenberg, An American couple, were executed in 1953 for passing atomic secrets to the Soviet Union.

355. The Aldrich Ames affair is one of the most notorious espionage cases in U.S. history. He was a CIA counterintelligence officer who spied for the Soviet Union and later Russia from 1985 until his arrest in 1994.

356. Pablo de la Guerra: The first Hispanic U.S. Marshal, appointed for the Southern District of California in 1850.

357. Thurgood Marshall was the first Black Supreme Court Justice, appointed by President Lyndon B. Johnson and sworn in on October 2, 1967.

358. U.S. Marshal Bass Reeves one of the first Black U.S. Marshals west of the Mississippi River tracked down his own son. Reeves' son, Bennie, was charged with the murder of his wife. Reeves took on the case tracked Bennie down, and brought him to justice.

359. In 1920, women had the right to vote once the 19th Amendment was ratified, giving.

360. The "Wedding Sting" refers to a clever police operation that took place in Owosso, Michigan, in 1990. The local police department, lacking the resources to arrest all the suspects individually, staged a fake wedding to gather them all in one place.

361. Anna Chapman: A Russian spy who was part of a sleeper cell in the United States. She was arrested in 2010 and later deported to Russia in a spy swap.

362. In 2015, a New York lawyer tried to settle a civil suit through a trial by combat!

363. Mel Carnahan, the former Governor of Missouri, was reelected to the U.S. Senate after he died in a plane crash in 2000.

12

CRIMES & OUTLAWS

364. The first train robbery in the United States occurred on October 6, 1866. The Reno brothers, John and Simeon, staged this historic heist by boarding an Ohio & Mississippi train shortly after it left Seymour, Indiana. They managed to steal $13,000 by breaking into one safe and tipping another off the train before making their escape.

365. The first bank robbery in the United States is a bit of a historical debate, but one of the earliest recorded bank robberies occurred on August 31, 1798, when a man named Isaac Davis and his accomplice, a bank porter, stole $162,821 from the Bank of Pennsylvania at Carpenters' Hall in Philadelphia.

366. They got away with it in the light of day. One of the most successful and infamous bank robberies in the Wild West was carried out by the James-Younger Gang. On February 13, 1866, they robbed the Clay County Savings Association in Liberty, Missouri. This heist is notable for being the first successful bank robbery in the light of day during peacetime in the United States. The gang escaped with approximately $60,000, a substantial amount at the time.

367. One of the most famous gangs was the Doolin Gang (the wild bunch) They were also known as the "Oklahoma Long Riders" because of the long dusters they wore. In the 1890s they robbed banks stores and trains and killed lawmen.

368. He jumped out of a perfectly good plane. D.B. Cooper is the alias of a man who hijacked a Boeing 727 on November 24, 1971. Cooper claimed to have a bomb and demanded $200,000 in ransom and four parachutes. After the plane landed in Seattle, he released the passengers when they brought him the money and parachutes, then instructed the crew to fly towards Mexico City. About 30 minutes after takeoff, Cooper opened the aircraft's aft door and parachuted into the night over southwestern Washington never to be found. The case remains the only unsolved air piracy incident in commercial aviation history.

369. "I've labored long and hard for bread, for honor and for riches, But on my corns too long you've tread, You fine-haired sons of? #@" Black Bart's poetic flair and gentlemanly demeanor made him a unique and memorable American outlaw robbing stagecoach in 1877. The verses he left behind added a touch of humor and sophistication to his otherwise criminal activities.

370. In 2011, Richard James robbed a bank for $1.00. It seems he wanted to go to federal prison for free health care.

371. Don't lend out stolen money. One of the most successful bank robberies in modern U.S. history was the Dunbar Armored robbery in 1997. A group of six men, led by Allen Pace, managed to steal $18.9 million from the Dunbar Armored facility in Los Angeles. They almost got away with it, but one of the robbers made a mistake by lending some of the stolen cash to a friend without removing the original cash straps, which led to their capture.

372. He hated being called Bugsy. The nickname originated from his violent temper and unpredictable personality, with people saying he was "crazy as a bedbug." Siegel preferred to be called "Ben" by his friends and "Mr. Siegel" by strangers. He is perhaps best known for his role in the development of Las Vegas. Although the initial opening was a disaster, the Flamingo eventually became successful and marked the beginning of the Las Vegas Strip. That was, of course, after he was killed.

373. Bonnie Parker and Clyde Barrow, often referred to as Bonnie and Clyde, were infamous American outlaws during the Great Depression. They became notorious for their bank robberies, small store heists, and murders across the Central United States. Their crime spree lasted from 1931 to 1934, capturing the public's imagination and making them media sensations.

374. Jean Terese Keating was involved in a fatal car crash in 1997, but she fled the state while awaiting trial in 1998. She was found in Canada in 2013 after bragging about getting away with the crime in a local bar.

375. It seems the powers that be at Alcatraz had a bit of a sadistic streak. They figured that if they gave the prisoners hot showers, they wouldn't get used to the frigid waters of San Francisco Bay.

376. Vlado Taneski was a Macedonian crime reporter who turned out to be a serial killer. Born in 1952, Taneski had a career in journalism spanning over 20 years. He wrote detailed articles about the murders of several women in his hometown of Kičevo, Macedonia. However, his articles contained information that had not been released to the public, which led to his arrest in June 2008.

377. The most famous escape from Alcatraz occurred in June 1962, involving three inmates: Frank Morris and brothers John and Clarence Anglin. They managed to break out of the maximum-security prison by creating dummy heads to fool the guards, digging through the walls, and constructing an improvised raft from raincoats. Some believe that Frank made it to shore.

378. Texas Seven Escape (2000): Seven inmates escaped from the John B. Connally Unit in Texas by overpowering prison workers, stealing uniforms and weapons, and fleeing in a stolen truck. They committed several crimes while on the run before being captured.

379. John Dillinger Escape (1934): The infamous bank robber escaped from the Lake County Jail in Indiana with a fake gun carved out of wood and used it to intimidate guards.

380. Minnesota sends Mosquitoes after escapees. In 2002, two inmates, James Robert Turner and Michael Wayne Murphy, escaped from the Faribault Correctional Facility in Minnesota. They managed to break out but were soon overwhelmed by the state's notorious mosquitoes. After just a few days on the run, they surrendered to authorities, citing the relentless mosquito bites as a major reason for giving up.

381. The cold case murder of Wayne Pratt, a gas station attendant in Winnebago County, Wisconsin. On June 12, 1963, Pratt was stabbed 53 times. For decades, the case remained unsolved. However, recent advancements in DNA technology have finally identified the suspect as William Doxtator. Unfortunately, he passed away, so no charges will be filed.

382. Genealogy and DNA Technology: In 2023, the Golden State Killer case was finally resolved using genealogy and DNA technology. a former police officer, Joseph James DeAngelo, was arrested and convicted in California for a series of murders as well as rapes during the 1970s and 1980s.

383. In 2022, after being on the run for over 40 years, William Bradford Bishop, who was accused of murdering his family in 1976, was found to be in a remote area in Mexico. Unfortunately, he had passed away before he could be brought to justice.

384. Christopher Knight, some called him the "North Pond Hermit." lived in the woods of Maine for 27 years, surviving by stealing food and supplies from nearby homes and campsites. He committed around 1,000 burglaries during this time. Knight was finally caught in 2013 while stealing from a camp for people with disabilities.

385. In the United States, a significant portion of violent crimes go unreported. According to recent data, only about 41.5% of violent crimes are said to have been reported to the police in 2022. This means that nearly 60% of violent crimes were not reported.

386. "Twinkie Defense" In 1978, Dan White used this defense after killing San Francisco Mayor George Moscone and Supervisor Harvey Milk. His lawyers argued that his consumption of junk food, including Twinkies, was evidence of his depression, which contributed to his diminished capacity. Surprisingly, this defense helped reduce his conviction to voluntary manslaughter.

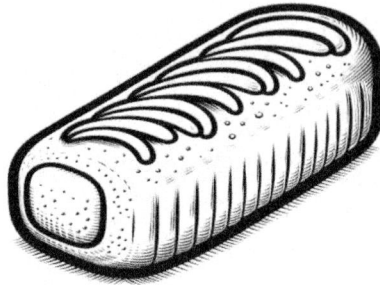

387. Many people are unaware that in the U.S, most burglaries occur during the daytime, specifically between 10 a.m. and 3 p.m. Burglars often target homes during these hours because they believe no one will be present.

388. John Wilkes Booth: After assassinating President Abraham Lincoln in 1865, Booth fled and was pursued by Union soldiers. He was eventually found and killed in a Virginia barn.

389. Ted Bundy, One of America's most notorious serial killers, escaped from custody twice in the 1970s. His final capture in Florida led to his conviction and execution.

390. In 2013, former LAPD officer Christopher Dorner went on a killing spree targeting law enforcement officers. A massive manhunt ended with his death in a cabin in Big Bear Lake, California.

391. Eric Rudolph: Known as the Olympic Park Bomber, Rudolph evaded capture for five years after the 1996 Atlanta bombing. He was finally arrested in 2003 while scavenging for food in North Carolina.

392. The Zodiac Killer: Police became aware of him in the late 1960s and early 1970s; He taunted police with cryptic letters and ciphers. Despite numerous suspects and extensive investigations, the killer's identity remains unknown.

393. In 1947, Elizabeth Short, known as the Black Dahlia, was found murdered in Los Angeles. Her case became one of the most famous unsolved murders in American history, with numerous theories but no definitive answers.

394. The Disappearance of Jimmy Hoffa: The former Teamsters Union leader vanished in 1975. Despite extensive searches and numerous theories, his fate remains a mystery.

395. The "Affluenza Defense": In 2013, Ethan Couch, a teenager, used this defense after causing a fatal drunk driving accident. His lawyers argued that his wealthy upbringing led to a lack of understanding of the consequences. He received probation instead of a prison sentence, sparking widespread outrage.

396. In 1987, Kenneth Parks was acquitted of murder after claiming he was sleepwalking when he killed his mother-in-law and attacked his father-in-law. The court accepted that he was unaware of his actions due to his sleepwalking state.

397. In 2005, a man named Ronald MacDonald (note the slightly different spelling) was arrested for stealing money from a Wendy's restaurant in Manchester, New Hampshire. He has no connection to McDonald's.

398. The Pendleton Correctional Facility in Indiana has a program called F.O.R.W.A.R.D. (Felines and Offenders Rehabilitation with Affection, Reformation, and Dedication). This program allows inmates to care for cats rescued from shelters. The program has been so successful that there's even a waiting list for inmates who want to participate.

399. Leon Gregg escaped from Georgia State Prison the night before his scheduled execution in July 1980. His freedom was short-lived. Just a few hours after his escape, he was killed in a bar fight in North Carolina.

400. In 1995, Erin Gilbert vanished from the Girdwood Forest Fair in Girdwood, Alaska. She was last seen with a man she had recently met, but no trace of her has ever been found despite numerous searches and investigations.

401. In 201 Michael LeMaitre disappeared while participating in the Mount Marathon Race in Seward, Alaska. Despite extensive searches, no trace of him has ever been found.

402. Sam Bass - Known for his failed train robberies, Bass was eventually betrayed by a member of his own gang and killed by Texas Rangers.

13

INTERESTING BUILDINGS

403. It's like a time capsule - The Wing Fort House is the oldest home in New England that's been continuously owned by the same family. Built in 1641.

404. The Fairbanks House in Dedham, Massachusetts, built between 1637 and 1641, was owned by the Fairbanks family for eight generations before becoming a historic museum.

405. The Brooklyn Bridge is an iconic suspension bridge connecting the boroughs of Manhattan and Brooklyn in New York City. It spans the East River and was completed in 1883, making it the longest suspension bridge in the world at the time.

406. The James Coolidge Octagon House in Madison, New York, was built around 1850, and it is believed to be the only existing octagonal cobblestone house in the world.

407. The camera loves her. The Carson Mansion (1886) - Located in Eureka, California, this Victorian house is considered one of the most photographed Victorian houses in the United States. Its intricate design and elaborate details make it a standout example of Queen Anne architecture.

408. The "Clinton's Ditch," as it was Nicknamed, after New York Governor DeWitt Clinton, who championed its construction. **The Erie Canal** was completed in 1825. This 363-mile-long canal was an engineering marvel connecting the Great Lakes to the Hudson River and New York City.

409. It's still used today. The San Miguel Chapel is a historic Spanish colonial mission church located in Santa Fe, New Mexico. It is often referred to as the oldest church in the United States, with its original construction dating back to around 1610. The church was built by Tlaxcalan Indians who accompanied Spanish settlers from Mexico.

410. It is still holding water. Old Oaken Bucket Pond Dam is located in Scituate, Massachusetts. It was constructed in 1640 and is still active today. This dam was originally built to support local mills and industries that required mechanical hydropower.

411. The Mark Twain House (1874) had a secret staircase from the first floor to the second floor. This hidden gem was only discovered during renovations in the 1960s.

412. Eads Bridge (1874) - The first steel arch bridge spans the Mississippi River at St. Louis and was a pioneering use of steel in bridge construction.

413. Very Corny! The Corn Palace (1892) Located in Mitchell, South Dakota, this unique building is decorated with corn and other grains.

414. The largest privately owned house in the U.S. is the Biltmore Estate (1895), Situated in Asheville, North Carolina. Built by George Washington Vanderbilt II, it has 250 rooms, including 35 bedrooms, 43 bathrooms, and 65 fireplaces.

415. Tuttle Farm (1632) - Situated in Dover, New Hampshire, this farm has been in the Tuttle family for nearly 400 years. It is one of the oldest continuously operating family farms in the United States.

416. How tall can we go? The Oroville Dam in California (1968) is the tallest earth-fill dam in the United States. In 2017, the dam made headlines when its spillway failed, leading to the evacuation of nearly 200,000 people.

417. That's a lot of concrete. Grand Coulee Dam (1942) - Situated on the Columbia River in Washington, it is the largest concrete structure in the U.S., containing a whopping 11.975 million cubic yards of concrete. That's enough concrete to build a sidewalk to the moon... well, almost.

418. Do you like going on picnics? This is the building for you. Located in Newark, Ohio, this building is shaped like a giant picnic basket. It was originally the headquarters of the Longaberger Basket.

419. Some still like books. The Kansas City Public Library - The facade of this library in Kansas City, Missouri, is designed to look like a giant bookshelf featuring classic book titles.

420. WonderWorks - This attraction in Pigeon Forge, Tennessee, is designed to look like an upside-down building, complete with inverted furniture and decor.

421. The development and use of autonomous construction equipment have been ongoing for more than 20 years, starting with the U.S. military.

422. 3D printing in construction is like a giant, concrete-spewing robot that builds houses and buildings layer by layer, like a 3D printer on steroids.

423. The 1st to stand tall. The Home Insurance Building in Chicago, completed in 1885, is often considered the world's first skyscraper. Designed by William Le Baron Jenney, this 10-story building was a pioneer in the architectural world, paving the way for the modern skyscrapers we see today.

424. Amazon doesn't sell this. The Sears Modern Homes were sold as kits from 1908 to 1940. Many Sears home kits are still standing today! Between 1908 and 1940, Sears sold around 70,000 to 75,000 kit homes across the United States.

425. The largest barn in America is a polo barn in Wellington, Florida. This impressive structure covers 78,000 square feet under its roof, with 62,267 square feet dedicated to the horse barn area.

426. One of the smallest houses in the USA is a tiny, two-story home called The Spite House, located in Alexandria, Virginia. Built in 1830 by John Hollensbury, this skinny structure measures just 7 feet wide and 25 feet deep. Hollensbury, tired of the noise and commotion caused by horse-drawn wagons and loiterers in the alley next to his home, decided to take matters into his own hands. He built this narrow house right in the middle of the alley, effectively blocking it off and putting an end to the nuisance.

427. Shirley Plantation (1638) - Located in Charles City, Virginia, this plantation has been owned and operated by the same family for eleven generations.

428. Do you believe in UFOs? The Spaceship House, also known as the Sculptured House or the Sleeper House, is a distinctive elliptical curved house built in Genesee, Jefferson County, Colorado, on Genesee Mountain in 1963 by the architect Charles Deaton. The house gained fame for its appearance in the 1973 science fiction comedy film "Sleeper" directed by Woody Allen, where it served as one of the main sets.

429. The world's largest tree house, known as the Minister's Treehouse, was in Crossville, Tennessee. It was built by a man named Horace Burgess, who claimed that God told him to build it. The structure was supported by an 80-foot-tall white oak tree and stood 97 feet tall, with 80 rooms spread across five stories. Unfortunately, the tree house burned down in October 2019.

430. When it was completed in 1937, the Golden Gate Bridge was, at the time, the longest and tallest suspension bridge in the world. Its main span stretches 4,200 feet (1,280 meters), and its towers rise 746 feet (227 meters) above the water.

431. The Golden Gate Bridge is painted with a specially formulated color known as International Orange. This paint is designed to protect the bridge from the harsh marine environment.

14

INTERESTING GEOGRAPHY

432. Imagine Alaska as the quirky cousin at the U.S. family reunion, showing up with an unexpected twist: it's both the farthest east and west in the country. Picture the Aleutian Islands stretching like a playful cat, with its tail, Attu Island, dipping into the Eastern Hemisphere. Meanwhile, the rest of Alaska lounges comfortably in the west, making it the only state where you can literally watch the sun set in the "east" over Russia. So, while other states might argue over who's more "out there," Alaska's just chilling at both ends of the map, proving once and for all that it's not just big; it's everywhere!

433. The modern concept of the prime meridian was established in 1884 at the International Meridian Conference in Washington, D.C. The conference selected the meridian passing through the Airy Transit Circle at the Royal Observatory in Greenwich, England, as the international standard.

434. The International Date Line (IDL) is an imaginary line that traverses from the North Pole all the way down to the South Pole, roughly following the 180-degree meridian of longitude. It serves as the boundary where each calendar day begins. When you cross the IDL from east to west, you add a day; when you cross from west to east, you subtract a day.

435. Ka Lae, also known as South Point, is on the Big Island of Hawaii. It's also the southernmost point in the United States. And you can dive off a 40-foot cliff while you are there.

436. It's lonely. Hawaii is the most isolated population center on Earth, located about 2,400 miles from California, 3,800 miles from Japan, and 4,900 miles from China.

437. Denali, located in Alaska, is the highest mountain in the United States. It stands at an impressive 20,310 feet above sea level.

438. First Ascent: Hudson Stuck, Harry Karstens, Walter Harper, and Robert Tatum made the first successful ascent of Denali in 1913. Harper, an Alaska Native, was the first to reach the summit.

439. Colorado has the most "14ers" (mountain peaks over 14,000 feet), with 53. Alaska comes in second with 29.

440. Known for his speed records, Andrew Hamilton climbed all 58 Colorado fourteeners in just 9 days, 21 hours, and 51 minutes in 2015. He also became the first person to summit all Colorado fourteeners in one winter.

441. Danelle Ballengee: A renowned endurance athlete, she holds the female-supported record for climbing all the Colorado fourteeners, completing them in 14 days, 14 hours, and 49 minutes in 2000.

442. Water at its freshest. The highest-named lake is in Colorado. It is Pacific Tarn, located in Summit County. It sits at an elevation of 13,420.

443. Still in Colorado. The highest reservoir in the United States is Lake Granby in Colorado. It sits at an elevation of 8,280 feet.

444. Want to climb a 14ner without climbing? Take the highest paved road in the United States to the top of Mount Evans in Colorado. This road reaches an elevation of 14,130 feet at its peak.

445. The highest continuous paved road in the United States is Trail Ridge Road in Colorado. This scenic route reaches an elevation of 12,183 feet and spans 48 miles through Rocky Mountain National Park, only open in the summer so you can drive from Estes Park and Grand Lake.

446. The largest cave in the United States is Mammoth Cave in Kentucky. It is also the longest cave system in the world, with over 426 miles (685.6 kilometers) of surveyed passageways.

447. The tallest tree in the world is Hyperion, a coastal redwood (Sequoia sempervirens) located in a remote area of Redwood National Park, California. Hyperion stands at an impressive 380.3 feet (115.92 meters) tall. The exact location of Hyperion is kept secret.

448. Crater Lake in Oregon Gets the distinction as the deepest lake in the United States. It reaches a depth of 1,949 feet. This stunning lake, formed in the caldera of a collapsed volcano.

449. Lake Superior is the largest lake in the United States. Its massive surface area of 31,700 square miles makes it not only the largest lake in the U.S. but also the largest freshwater lake in the world by surface area.

450. The Blue Earth River in Minnesota has been known to flow backward during extreme flooding events. This unusual phenomenon occurs when heavy rainfall causes the river to swell and reverse its flow direction.

451. The largest boulder in the USA is known as Giant Rock and can be found in the Mojave Desert, California; this massive freestanding boulder covers 5,800 square feet and stands seven stories high. It's considered the largest freestanding boulder in North America and is even purported to be the largest in the world.

452. It blew its top, literally. Mount St. Helens! The volcano decided to have a major meltdown in 1980. The eruption blew off the north side of the mountain. This event was one of the largest and most destructive volcanic eruptions in U.S. history, Shortening the mountain by about 1,300 feet!

453. The largest snow slide in the United States occurred in 1910 and is known as the Wellington avalanche. This tragic event took place near Stevens Pass in the Cascade Mountains of Washington state. A heavy snowfall and a thunderstorm triggered a massive snow slide that swept two trains off the tracks and into a ravine, killing 96 people. It was the deadliest avalanche in US history.

454. In 1856, New York initially allocated $5 million to purchase the land for Central Park. However, by the time it was completed, the total cost of designing and constructing the park reached around $14 million.

455. Grand Falls, also known as Chocolate Falls, is a spectacular waterfall located in the Painted Desert on the Navajo Nation, about 30 miles northeast of Flagstaff, Arizona. At 181 feet tall, Grand Falls is taller than Niagara Falls, which is 176 feet.

456. Reelfoot Lake in Tennessee was created by a series of powerful earthquakes known as the New Madrid earthquakes, which occurred between December 1811 and February 1812. These earthquakes were so intense that they temporarily caused the Mississippi River to flow backward, leading to the formation of the lake.

15

ANIMALS

457. " The black-footed ferret (Mustela nigripes) is a North American mammal that was once thought to be extinct. However, in 1981, a small population of black-footed ferrets was discovered in Wyoming, USA. Since then, conservation efforts have been made to save this species from extinction.

458. Lincoln Park Zoo's first animal acquisition was a bear cub from the Philadelphia Zoo in 1874. This little cub was purchased for the grand sum of $10, marking the beginning of the zoo's animal collection.

459. One of the biggest trout caught in recent memory was a 32-inch tiger trout caught at Scofield Reservoir in Carbon County, Utah.

460. Is Bigfoot real? In Skamania County, Washington, it's illegal to kill Bigfoot. This law was first enacted in 1969 due to the surge in Bigfoot sightings. The county even declared itself a "Sasquatch Refuge".

461. The kangaroo rat in Death Vally can live its entire life without drinking a single drop of liquid. This little rodent gets all the moisture it needs from the seeds it eats.

462. Flamingos eat with their heads upside down, using their specially adapted bills to filter out food from the water.

463. Honeybees are not native to the US! They were brought over by European settlers in the 17th century.

464. Roadrunners can reach speeds of up to 25 miles per hour, making them one of the fastest-running birds.

465. Often considered the deadliest snake in North America: The Mojave rattlesnake (Crotalus scutulatus). It is found in the deserts of the southwestern United States and parts of Mexico, making encounters with humans relatively common.

466. The Gila monster is the only venomous lizard native to the United States. The Gila monster's bite is venomous, but it is rarely fatal to humans.

467. Praying mantises are the only insects known to have stereo vision, which means they can perceive depth and see in 3D.

468. Prairie dogs live in extensive underground burrows called "towns," which can cover hundreds of acres. These burrows have designated areas for nurseries, sleeping, and even toilets.

469. Florida is the only place in the world where you can find alligators and crocodiles coexisting in the wild.

470. The largest land animal in the United States is the American bison. Male bison, also known as bulls, can weigh up to 2,000 pounds and stand about 6 feet tall at the shoulder.

471. Male American alligators can grow up to 15 feet long and weigh as much as 1,000 pounds. Females are generally smaller, averaging around 8.2 feet in length. Found only in America.

472. Gunnison Sage-Grouse: Native to Colorado and Utah, this bird is known for its elaborate courtship rituals.

473. Hellbender: This giant salamander lives in the rivers and streams of the Appalachian Mountains.

474. Red Wolf: Once widespread, the red wolf is now found only in certain parts of the southeastern United States.

475. Donkeys have a unique visual capability due to the placement of their eyes on the sides of their heads. This positioning gives them nearly 360 degrees around them. However, it's a bit of a myth that they can see all four of their feet simultaneously.

476. At Utah State University, researchers led by Professor Randy Lewis have implanted spider genes into goats. These genetically modified goats produce milk containing spider silk proteins. The silk proteins are then extracted from the milk and spun into spider silk thread, which is incredibly strong and flexible.

477. Researchers have created glow-in-the-dark cats to study diseases like HIV/AIDS. Fluorescent protein helps track the spread of the virus in the body.

478. Faith is the incredible two-legged dog who became an inspiration to military personnel. Faith was born with only two functional legs and learned to walk upright like a human. She and her owner, Jude Stringfellow, visited numerous military hospitals and bases, bringing hope and joy to wounded soldiers.

479. It's the state's official reptile. The Alabama Red-bellied Turtle, This turtle with its red belly can grow up to 16 inches and is found only in the Mobile-Tensaw River Delta in Alabama.

480. Alabama Sturgeon: This critically endangered fish is found only in the Mobile River Basin. Known for its distinctive yellowish-orange color, it grows to about 30 inches long.

481. The Salmon Cannon, also known as the Whooshh Fish Transport System, is a fish migration tool designed to help salmon and other migratory fish overcome barriers like dams to reach their spawning grounds.

482. It's like a vampire but for bees. The Varroa mite, also known as Varroa destructor, is a tiny, reddish-brown parasite that sucks the blood of honeybees, leading to weakened and diseased colonies. They have found a way to breed bees resistant to these vampires.

483. One unique animal found only in Alaska is the Alexander Archipelago Wolf. This rare subspecies of the gray wolf lives exclusively on the islands in the Alexander Archipelago and a small strip of coastline separated from the mainland by mountains.

484. Texas is the state with the most cows in the United States. It boasts an impressive population of around 4.3 million beef cows, which accounts for nearly 15% of all beef cows in the country.

485. Texas has the largest number of bats in America. The state is home to the Bracken Cave Preserve, which hosts over 15 million Mexican free-tailed bats during the summer, making it the world's largest bat colony.

486. The largest bat in the United States is the mastiff bat (Eumops perotis). These big boys can have a wingspan of over 20 inches and weigh up to 2.2 ounces.

487. In 1925, a diphtheria outbreak threatened the remote town of Nome, Alaska. The only way to get life-saving serum to the town was by dog sled. The journey covered over 600 miles in brutal weather conditions, with temperatures dropping to -60°F (-51°C) and winds reaching hurricane force. The final leg of the journey was completed by Gunnar Kaasen and his lead dog Balto, who delivered the serum to Nome, saving countless lives.

488. In the late 1960s, Fu Manchu, the orangutan at Omaha's Henry Doorly Zoo (Nebraska), repeatedly escaped his enclosure by using a piece of wire to pick the lock. He would hide the wire between his lip and gums, taking it out only when he wanted to make his escape. He never caused trouble; he enjoyed exploring other parts of the zoo.

489. Louisiana is responsible for about 90% of the harvested in the United States. The state has a thriving crawfish industry, with over 250,000 acres of ponds and over 1,600 farmers.

16

TRANSPORTATION

490. The first stop sign in the United States was installed in 1915 in Detroit, Michigan.

491. The original stop sign was not our current octagonal shape. It was square and white with black letters spelling out "STOP." It was smaller than modern stop signs, reflecting the lower speeds of early automobiles and less crowded streets.

492. The first state to issue license plates for cars, starting in 1903, was Massachusetts.

493. In 1923, the Mississippi Valley Association of State Highway Departments recommended the use of octagonal stop signs.

494. The first police car was actually an electric wagon used by the Akron Police Department in Akron, Ohio. It was introduced in 1899 and could not go faster than 16 mph. It had a range of 30 miles.

495. The first electric traffic signal in the United States was installed on August 5, 1914, at the intersection of Euclid Avenue and East 105th Street in Cleveland, Ohio. This early traffic light system featured red and green lights to control traffic flow.

496. Slow down!! Connecticut's Automobile Law (1901) Speed Limits: The law sets a maximum speed limit of 12 miles per hour while in the city and 15 miles per hour once you leave the city.

497. In 1901, New York registration vehicles cost $1.00, and if you were driving your auto for hire, you had to Pay $1.00 for an operator's (chauffeur) license.

498. Jean-Pierre Blanchard is credited with the first hot air balloon flight in the United States on January 9, 1793. Blanchard ascended from the Washington Prison Yard in Philadelphia and landed in Gloucester County, New Jersey. This historic flight was witnessed by President George Washington, which helped spur interest in ballooning in the United States.

499. In 1903, Horatio Nelson Jackson and Sewall Crocker completed the first successful road trip across the United States in an automobile. They left San Francisco on May 23 and arrived in New York City on July 26, a journey that took 63 days. They were driving a 1903 Winton touring car.

500. In 1891, one of the first recorded accidents involving a gasoline-powered buggy occurred in Ohio City, Ohio. The driver was James Lambert, who lost control and crashed into a hitching post.

501. The first recorded car accident in the United States involving another person occurred on May 30, 1896, in New York City. The accident involved Henry Wells, who was driving a Duryea Motor Wagon. Henry Wells collided with a bicyclist named Evelyn Thomas. Thomas sustained a broken leg, marking the first recorded injury in a car accident in the United States.

502. This tragic incident led to the first recorded lawsuits due to a car accident in the United States.

503. Henry Ford's introduction of assembly-line production in 1913 made personal transportation more accessible. Ford's Model T revolutionized the automobile industry and made cars affordable for the average person.

504. One odd fact about the Model T car is that it had no gas pump in the tank. Instead, the fuel is moved by gravity. This design quirk meant that if the car were driven up a steep hill, the gas would flow away from the carburetor, causing the engine to sputter. To solve this problem, drivers often had to turn the car around and drive up the hill in reverse.

505. Stanley Steamer (1902): This steam-powered car was known for its speed and smooth ride. The Stanley brothers' vehicles were among the fastest of their time, with some models reaching speeds of 127 mph.

506. The first official car race in the United States took place on Thanksgiving Day, November 28, 1895. This historic event, known as the Chicago Times-Herald race, covered a distance of 54 miles from Chicago to Evanston and back. The average speed is a little over 5 MPH.

507. A few years after the first official car race, American billionaire William K. Vanderbilt Jr. set a speed record of 111.8 km/h (69.5 mph) in 1902 in a Mercedes-Simplex.

508. Currently (2024), the fastest production car in the world is the SSC Tuatara. This hypercar, produced by SSC North America, set a new world record with a top speed of 316.11 mph. What street can you drive that on?

509. The first motorcycle in the United States was the Orient, a gasoline-powered bicycle produced by the Waltham Manufacturing Company in 1899.

510. 1936 - The first Harley Davidson knucklehead Motorcycle. This was the first new motorcycle during the Great Depression and was a bit of a gamble for Harley-Davidson.

511. The Wright brothers, Orville and Wilbur Wright, got off the ground for the first powered flight on December 17, 1903

512. Flying in a plane wasn't around long before someone had to jump out of one. Captain Albert Berry is the first person credited with making a successful parachute jump from a powered airplane. He made this historic jump on March 1, 1912, from a Benoist pusher biplane over St. Louis, Missouri.

513. The first commercial airplane was the Boeing 247, which made its debut in 1933 and could carry 10 passengers along with a crew of three (pilot, co-pilot, and stewardess). When it was first introduced, the cost to fly on the Boeing 247 varied, but a typical fare for a cross-country flight was around $160 one way.

514. The Boeing 247 was faster than many military aircraft of its time, with a cruising speed of about 189 mph.

515. The New York City Subway, which opened in 1904, became one of the largest and most extensive subway systems in the world. It took 4 years to build. The fair was only $.05, and it stayed at that price for 40 years.

516. A 3-speed bicycle known as "The Hill-Climber" was invented by Peter J. Scharbach in 1902.

517. It's crazy to think, but the first designated bike lane in the United States was established on Ocean Parkway in Brooklyn, New York, on June 15, 1894.

518. The first train to cross the Rocky Mountains was part of the transcontinental railroad, which was completed on May 10, 1869. The railroad was built by two main companies that met at Promontory Summit, Utah, where the "Golden Spike" was driven to commemorate its completion.

519. Early trains on the transcontinental railroad typically traveled at an average speed of about 20-25 miles per hour.

520. The Pan-American Highway is an impressive network of roads stretching from Prudhoe Bay, Alaska, to Ushuaia, Argentina, covering about 30,000 miles (48,000 kilometers). It's recognized by the Guinness World Records as the world's longest "motorable road". The highway runs through 14 countries, including the United States.

521. The United States does indeed boast the longest railway system in the world, stretching over an impressive 220,044 km. That's a lot of track to lay down, and it's certainly a feat of engineering!

522. The Ford Pinto, a subcompact car introduced in 1971, became notorious for its tendency to explode in rear-end collisions due to a flawed fuel tank design. The fuel tank was positioned in a vulnerable location, and a simple $11 part could have prevented the explosions.

17

POP CULTURE

523. The movie Don Juan, premiering on August 6, 1926, was the first to feature synchronized sound. It included a synchronized musical score and sound effects but no spoken dialogue.

524. November 18, 1928– Mickey Mouse appears in Steamboat Willie, the third Mickey Mouse cartoon released and the first animated film with synchronized sound (music and sound effects).

525. The first commercial radio station in the United States was KDKA in Pittsburgh, Pennsylvania. It began broadcasting on November 2, 1920 KDKA's first broadcast was due to the 1920 presidential election, where Warren G. Harding defeated James M. Cox2. This event marked the beginning of radio as a medium for mass communication in the U.S.

526. Why is there a billboard in Antarctica advertising a store in South Dakota? It's true! Wall Drug Store in Wall, South Dakota, is famous for its extensive and quirky billboard advertising campaign. The Wall Drug Store opened in December 1931. Ted Hustead and his wife Dorothy bought the only drugstore in Wall, South Dakota, and turned it into a famous roadside attraction with their clever advertising for free ice water. Today, Wall Drug's billboards can be found all over the world.

527. Annette Funicello, one of the original Mouseketeers and a beloved star of the 1960s "Beach Party" movies, was asked by Walt Disney not to show her belly button in her films. Despite the beach settings and her bikinis, Annette adhered to this guideline in most of her movies. It was a way to maintain a wholesome image, which was very important to Disney and aligned with the family-friendly brand he was building.

528. John Wayne was an avid chess player. He often carried a miniature chessboard with him. He often played with other Hollywood stars like Marlene Dietrich, Rock Hudson, and Robert Mitchum.

529. John Wayne was nicknamed "Duke" after his childhood dog, an Airedale Terrier named Duke. Locals started calling him "Little Duke," and the name stuck.

530. Interesting eating habit. Marilyn Monroe enjoyed eating raw carrots with her meals and often broiled her own steak, liver, or lamb chops in her hotel room.

531. Jennifer Lawrence created a "chili pizza sandwich," which involves putting chili and noodles between two slices of pizza.

532. Ed Sheeran travels with his own supply of ketchup to ensure he always has his favorite condiment.

533. Taylor Swift has been known to bring her own food to events and even eat from Tupperware at fancy gatherings.

534. Twain, the famous writer, was born shortly after Halley's Comet passed by Earth in 1835 and famously predicted he would die with its return. Remarkably, he passed away on April 21, 1910, the day after the comet's closest approach to Earth

535. Gunsmoke ran for a whopping 20 seasons, from 1955 to 1975, making it one of the longest-running prime-time TV series in the U.S.

536. On Gunsmoke, shootings were quite frequent, given the show's setting in the rough-and-tumble Dodge City. Marshal Matt Dillon, over the 20 years of the series, was shot 56 times.

537. Morris the Cat is a famous advertising mascot for the 9Lives brand of cat food. Morris was saved from a Chicago-area animal shelter in 1968. From 1969 to 1978, Morris starred in 58 television commercials.

538. The Pet Rock was quite a craze back in the mid-1970s. A rock packaged in a cardboard box with air holes and straw bedding was sold as a "pet." The man who came up with the idea was advertising executive Gary Dahl. Pet Rock became a cultural phenomenon, selling over a million units and making Dahl a millionaire.

539. The term "action figure" was first used in 1964 by the Hasbro Company's Don Levine to describe their new G.I. Joe toy.

540. Walt Disney claims the record for the most Oscars won by an individual, having received 22 competitive awards and 4 honorary awards.

541. For films, three movies share the record for the most Oscars won, each with 11 awards: "Ben-Hur" (1959) -"Titanic" (1997), -and "The Lord of the Rings: The Return of the King" (2003).

542. Johnny Depp recently turned to tarot iconography to inspire his latest art series, which includes pieces like "The Lovers," "The Empress," "The Emperor," and "Strength".

543. The New Hampshire state logo is "Live Free or Die".

544. "The Day the Music Died" refers to February 3, 1959, a plane crash near Clear Lake, Iowa, which took the lives of Buddy Holly, Ritchie Valens, and J. P. "The Big Bopper" Richardson, all American rock and roll musicians. This tragic event was immortalized in Don McLean's 1971 song "American Pie."

545. Mr. Beast is the top dog on YouTubers these days. With over 300 million subscribers, he's the king of the platform. He's building an empire of crazy challenges and philanthropy.

546. Microcelebrities refer to influencers, content creators, TikTok stars, and Instagram models who've managed to carve out a niche following in the vast wasteland of the internet.

547. Example of a Microcelebrity: The "Chewbacca Mom" - Candace Payne became an overnight sensation after her Facebook Live video of her laughing hysterically while wearing a Chewbacca mask went viral.

548. YouTube was officially started in February 2005 by Chad Hurley, Steve Chen, and Jawed Karim. They had difficulty sharing videos shot at a dinner party.

549. The first video uploaded to well-known YouTube was titled "Me at the Zoo," featuring Jawed Karim at the San Diego Zoo.

550. During their time on "The All-New Mickey Mouse Club," Justin Timberlake's mother, Lynn Harless, became Ryan Gosling's legal guardian for about six months.

551. The song "Rudolph the Red-Nosed Reindeer" has a heartwarming origin story. It all began in 1939 when Robert L. May, a copywriter for the Montgomery Ward department store, was given the job of creating a Christmas story to attract customers. May used his childhood experiences of being bullied to help create the character Rudolph.

552. In the 1990s, Michael Jackson tried to buy Marvel Comics. He was reportedly a huge fan of Spider-Man and wanted to produce and potentially star in a Spider-Man movie.

553. Sylvester Stallone had to sell his dog, Butkus, when he was struggling financially before his big break with "Rocky." He sold Butkus for $50 to a man outside a liquor store, but later repurchased him for a much higher amount after selling the "Rocky" script. Butkus was his dog in the movie.

554. Eddie Murphy is the only person who has hosted Saturday Night Live while still being a regular cast member.

18

CULTURAL ICONS

———❦———

Found more fun information about famous people, so had to share.

555. American Celebrities can seem self-serving. When an American catastrophe happens, Many ask the public to help. But some don't ask others to help; they jump in and start helping. One such celebrity is Actor Steve Buscemi. He was a firefighter before becoming an actor. When 911 happened, he went to New York, put on a spare firefighter uniform, and worked with the other firefighters on 12-hour shifts. He refused any interviews or publicity and just worked tirelessly to help. So if you see him, thank him.

556. Walt Disney was fired from a newspaper for "lacking imagination and having no good ideas."

557. "Friendly Neighborhood Spider-Man." This series has been published in various runs, with the most recent starting in 2019. The title is derived from Spider-Man's famous self-referential comment, "Just another service provided by your friendly neighborhood Spider-Man!"

558. One of the most notable collaborations with Johny Depp was when he joined Jeff Beck on stage in 2022 for a surprise performance in Sheffield, England. They played a cover of John Lennon's "Isolation,".

559. A Florida judge once labeled Elvis Presley, a "savage" for his supposed negative influence on the youth.

560. Jackie Chan has been blacklisted by insurance companies when it comes to his stunts.

561. Hattie McDaniel became the first Black woman to win an Academy Award. In 1940, she won the Oscar for Best Supporting Actress for playing Mammy in Gone with the Wind.

562. Dolly Parton once entered a Dolly Parton look-alike contest and, believe it or not, she lost! She exaggerated her look with bigger eyes, bigger hair, and a bigger beauty mark, but she lost to a man (A drag queen).

563. The first "talkie" was The Jazz Singer, released in 1927. Al Jolson was the main actor in this groundbreaking film. He played Jakie Rabinowitz, a young man who defied his family's traditions to become a jazz singer.

564. FTD (Florists' Transworld Delivery) delivered the most flowers in one day on August 16, 1977, the day Elvis Presley passed away.

565. The Princess and the Frog. After the film's release in 2009, there was a spike in salmonella cases among children. This was because many kids tried to imitate the movie's main character, Tiana, by kissing frogs. Over 50 children were hospitalized due to salmonella infections from kissing frogs.

566. The first film to have a budget of $100 million was "True Lies" (1994), directed by James Cameron and starring Arnold Schwarzenegger and Jamie Lee Curtis.

567. Alfred Hitchcock and his fear of eggs! Yes, that's right. The master of suspense had a phobia of eggs.

568. Johnny Depp - The actor has a fear of clowns (coulrophobia)

569. Billy Bob Thornton - He has multiple phobias, including a fear of antique furniture and clowns.

570. Matthew McConaughey has a fear of revolving doors.

571. Not only a legendary actor but also a successful race car driver, Paul Newman competed in professional racing and even won several championships.

572. Steve McQueen was Known as the "King of Cool" and was also a great racer, participating in events like the 12 Hours of Sebring.

573. James Garner was a skilled driver and raced in events like the Baja 1000.

574. Ronda Rousey had major roles in Furious 7 and The Expendables 31 was A former UFC Women's Bantamweight Champion and Olympic bronze medalist in judo.

575. Gina Carano - A pioneer in women's MMA, also became an actress and starred in Haywire and Deadpool.

576. Before transitioning into politics, Ronald Reagan had a prolific acting career, appearing in 53 feature films.

577. Hulk Hogan, before he was body-slamming giants and ripping his shirt off in the ring, Hulkster was swinging a mean bat and throwing some serious heat on the baseball diamond.

578. Walt Disney did not graduate from high school or attend college. He left high school at the age of 16 to join the Red Cross as an ambulance driver during World War I.

579. Steven Spielberg's big break came with a short film titled "Amblin'" (1968), which caught the attention of an executive at Universal Studios. This made Spielberg the youngest director to sign a long-term deal with a Hollywood studio.

580. Russell Brand is certainly a unique character with many odd and interesting aspects to his life. Despite his wild persona, Brand is a dedicated yoga practitioner. He even released a book titled "Recovery: Freedom from Our Addictions," where he discusses how yoga and meditation helped him overcome his struggles.

581. Joe Rogan is a multifaceted personality known for his work as a podcaster, comedian, UFC commentator, and actor. Launched in 2009, "The Joe Rogan Experience" podcast has become one of the most popular in the world.

582. Jackie Chan is more than an action star; he's also a singer. He's released over 20 albums and has sung in languages including Cantonese, Mandarin, English, and Japanese. He even won a Best Foreigner Singer Award in Japan in 1984.

583. As of 2024, Margot Robbie is the highest-paid female actress. She is expected to earn a staggering $50 million this year. Robbie is known for her versatile roles in films such as "The Wolf of Wall Street" and "Suicide Squad".

584. Dwayne "The Rock" Johnson has limited Sleep: Despite his intense schedule, he only gets about three to five hours of sleep each night.

585. Oh, Chloe Ting, the queen of sweat and crunches! With over 25 million loyal subjects in her YouTube kingdom. Her workout challenges are like royal decrees, commanding us to sweat and tone our way to fitness glory.

586. The traditional names for the wise monkeys are Mizaru, Kikazaru, and Iwazaru, which translate to "See no evil," "Hear no evil," and "Speak no evil," respectively.

587. Carmen Dell'Orefice is still modeling! At the age of 93, she continues to grace the covers and runways, proving that age is just a number in the world of fashion. The woman has been gracing the covers of Vogue since 1946.

19

AMERICAN CUISINE

---

588. In 2023, the United States sold a staggering 1.7 billion chicken sandwiches. That's enough to give every person in the country about five sandwiches each.

589. The first fast food restaurant is widely considered to be White Castle, which opened in 1921 in Wichita, Kansas. Sold hamburgers, which were sold for just five cents each.

590. Colonel Harland Sanders started his franchise in 1952. He franchised his secret recipe to Pete Harman (a friend), who operated one of the largest restaurants in South Salt Lake, Utah. Kentucky Fried Chicken was born.

591. When it opened in 1940 in San Bernardino, California, McDonald's was originally a barbecue restaurant. It wasn't until after World War II that the McDonald brothers shifted their focus to hamburgers.

592. The only state that grows coffee beans is Hawaii, although California produces a small amount.

593. Charles Taylor invented the first soft ice cream machine in 1926 in Buffalo, New York.

594. Glen Bell started Taco Bell in 1962 in Downey, California. Before that, he ran a series of hamburgers and hot dog stands.

595. George Crum, a chef in New York in 1853, was trying to appease a customer who kept sending back his fried potatoes, complaining they were too thick. In frustration, Crum sliced the potatoes thin, fried them until they were crispy, and added extra salt. To his surprise, the customer loved them, and thus, the potato chip was born. Fun but maybe not true????? Because of the following.

596. The earliest known recipe for something like potato chips dates to 1817 in England, found in William Kitchiner's cookbook, "The Cook's Oracle".

597. Fortune cookies were invented in the US.

598. Kool-Aid became the official state soft drink of Nebraska, USA in 1998.

599. After he sold the company in 1964. Colonel Harland Sanders often criticized the quality of the food, particularly the gravy, which he famously described as "wallpaper paste".

600. Carl's Jr, Carl Karcher, and his wife started with a hot dog cart in Los Angeles in 1941. They eventually opened Carl's Drive-In Barbecue, which became known for its hamburgers.

601. Coca-Cola was first sold as a medicine in 1886 by John Pemberton, a pharmacist. The original formula contained cocaine from the coca leaf and caffeine from the kola nut, which gave the drink its name. It was advertised as a "brain tonic" and "nerve stimulant" that could cure headaches and relieve exhaustion.

602. Buttermilk is the liquid left over after churning butter from cream (at least it used to be). Today, most buttermilk is cultured, meaning it's made by adding bacterial cultures to regular milk, which ferments it and gives it that tangy flavor.

603. In 1625, Reverend William Blaxton started the first apple orchard in New England. Apples are not native to North America.

604. Ruth Graves Wakefield made a deal with Nestlé. In 1939, she agreed to allow Nestlé to print her Toll House Cookie recipe on their packages of semi-sweet chocolate. In exchange, she received a lifetime supply of Nestlé chocolate.

605. In 1905, 11-year-old Frank Epperson accidentally left a mixture of powdered soda, water, and a stirring stick outside on a cold night. The next morning, he discovered it had frozen, creating the first popsicle and calling it "Epsicle."

606. Dr. John Harvey Kellogg, along with his brother Will Keith Kellogg, accidentally created cornflakes in 1894 while trying to make a new type of granola. They left cooked wheat out and it went stale. Instead of throwing it away, they rolled it out and toasted it, creating the first cornflakes.

607. Worcestershire Sauce was an accident created in the early 19th century by chemists John Lea and William Perrins. They were trying to recreate a sauce from India but found the initial batch inedible. They left it in the basement and forgot about it. Months later, they discovered it had fermented into a delicious sauce.

608. National Peanut Butter and Jelly Day is celebrated on April 2nd each year. Julia Davis Chandler published the recipe for a peanut butter and jelly sandwich in the "Boston Cooking School Magazine of Culinary Science and Domestic Economics" in 1901.

609. The well-loved grilled cheese became popular in the United States in the early 1900s. In 1928 Otto Frederick Rohwedder invented sliced bread, and James L. Kraft developed processed cheese in 1914, making it easier to prepare.

610. Turophobia is the irrational fear of cheese.

611. The Reuben sandwich! A classic American dish that's so good, it's practically a national treasure, with layers of corned beef, Swiss cheese, sauerkraut, and Russian dressing all piled high on rye bread. And let's not forget the crispy, buttery grilled exterior that adds a whole new dimension of deliciousness.

612. Philly Cheesesteak: Pat and Harry Olivieri, hailing from Philadelphia (1930), this sandwich is made with thinly sliced beefsteak and melted cheese, often topped with onions and served in a long hoagie roll.

613. In 1867, Charles Feltman, a German immigrant, had a hot dog-shaped epiphany. He started selling sausages in milk rolls from a pushcart in New York City's Bowery. And thus, the hot dog was born!

614. Milton S. Hershey founded the town of Hershey, Pennsylvania, in 1903. He built his famous chocolate factory there, and the town was developed to provide a pleasant living environment for his employees, complete with all the modern amenities, like electricity and indoor plumbing.

615. While some people simply dislike chocolate, those with xocolatophobia experience significant anxiety at the thought of consuming or even being near chocolate.

616. In the United States, approximately 8.2 million pizzas are eaten every day. This translates to around 75 million pounds of pizza eaten every day.

617. On average, U.S. households spend about $3,639 annually on eating out.

618. Harry Burnett Reese started out working for the Hershey Chocolate Company, and invented Reese's Peanut Butter Cups in 1928.

619. Skittles contain titanium dioxide. It's like the secret ingredient that gives them their vibrant colors.

620. Not on the menu but you can order it. The McGangBang: This is a McChicken sandwich placed inside a double cheeseburger. It's a hearty combination for those who want a bit of everything.

621. If you like sushi and have money to spare, the Masa (New York, NY) sushi restaurant offers an omakase dining experience with prices around $600 per person.

622. New Jersey is known as the diner capital of the world. They claim to have more than 500 diners spread across the state.

623. The term "cheeseburger" was trademarked in 1935 by Louis Ballast of the Humpty Dumpty Drive-In in Denver, Colorado.

624. Around 1890, Evanston, Illinois, passed a law prohibiting the sale of soda on Sundays. To circumvent this, soda fountains began serving ice cream with syrup instead of soda, and that's how we got the ice cream sundae.

625. The first potatoes in the United States were planted in 1719 by Scotch-Irish immigrants in Londonderry, New Hampshire.

626. According to the latest data, mayo is raking in a whopping $2 billion annually, leaving ketchup in the dust with a measly $800 million.

627. During World War II, the U.S. government temporarily banned the sale of sliced bread on January 18, 1943, as part of a broader effort to conserve resources.

20

SPORTS

628. Too many home runs? The thin air at Coors Field, home of the Colorado Rockies, makes it easier to hit a home run. The high altitude reduces air resistance, allowing the ball to travel further. It's like a hitter's dream come true!

629. Since 2002, baseballs at Coors Field have been stored in a humidor. This keeps the balls at a consistent humidity level, making them less bouncy and reducing their travel distance.

630. In 2016, the Rockies added higher fences in the outfield, specifically an 8.75-foot-high chain-link fence from right-center to right field. This makes it harder for balls to clear the wall.

631. Lefty. Babe Ruth - Known as "The Sultan of Swat," Ruth's 1921 totals of 177 runs, 457 total bases, and 119 extra-base hits all remain modern-era major league records Babe Ruth is arguably the most iconic figure in baseball history.

632. Alexander Cartwright Often credited with formalizing the modern rules of baseball. In 1845, he established the Knickerbocker Rules, which included the diamond-shaped infield and the three-strike rule.

633. The first recorded baseball game under Cartwright's rules was played in 1846 between the Knickerbocker Club and a team of cricket players.

634. In December 1891, James Naismith decided to hang a couple of peach baskets at each end of a gymnasium, and voila! Basketball was born The first public game was played in Springfield, Massachusetts, on March 12, 1892.

635. The first Super Bowl was played on January 15, 1967, at the Los Angeles Memorial Coliseum in California. This historic game featured the NFL-champion Green Bay Packers against the AFL-champion Kansas City Chiefs. The Green Bay Packers emerged victorious, defeating the Kansas City Chiefs with a score of 35-10

636. On Super Bowl Sunday, pizza orders can increase by up to 40% compared to a regular Sunday. An estimated 12.5 million pizzas are ordered on this day.

637. German immigrant, Chris Von der Ahe noticed that baseball games brought more business to his saloon, so he had the genius idea to introduce hot dogs to the crowds at the game.

638. Chuck Connors. Before becoming famous for his role as Lucas McCain in the TV series "The Rifleman," he had a career in Major League Baseball (MLB) and professional basketball. He played for the Brooklyn Dodgers and the Chicago Cubs in the MLB.

639. American football is the most popular sport in the United States. It has a massive following, and the NFL (National Football League) is the most-watched sports league in the country.

640. Tug of War was last included in the Olympics in 1920,

641. Jousting is the official state sport of Maryland. It was designated as such in 1962, making Maryland the first state to adopt an official sport.

642. Christian Coleman got some serious speed, holding personal bests of 9.76 seconds for the 100 meters and 19.85 seconds for the 200 meters. He's also the world record holder for the indoor 60 meters with a blazing time of 6.34 seconds.

643. The longest professional baseball game ever played was between the Pawtucket Red Sox and the Rochester Red Wings. This marathon game lasted 33 innings and took a total of 8 hours and 25 minutes to complete. It was finally completed on June 23, 1981, with Pawtucket winning 3-2.

644. The lowest-scoring professional basketball game in NBA history took place on November 22, 1950, between the Fort Wayne Pistons and the Minneapolis Lakers. The final score was 19-18 in favor of the Pistons, with a combined total of just 37 points.

645. A famous low-scoring game was the 0-0 tie between the Detroit Red Wings and the Montreal Maroons on March 24, 1936. This game went into six overtime before Mud Bruneteau. scored the only goal, ending the longest game in NHL (National Hockey League) history.

646. Golf was introduced to America by Scottish and English immigrants in the late 17th century. The first recorded game took place in 1779 on a New York City farm.

647. The United States Golf Association (USGA) was founded in 1894 to standardize rules and organize national championships.

648. Alaska's official state sport is dog mushing, also known as dog sledding. It was designated as such in 1972.

649. Twiggy the Water-Skiing Squirrel has been entertaining audiences since 1979. What started as a joke about teaching a squirrel to water ski turned into a full-fledged act that has spanned multiple generations of squirrels.

650. In Maine, there's an annual event called the Damariscotta Pumpkinfest & Regatta, where participants carve out giant pumpkins and race them in the water.

651. Golf is the only sport to have been played on the moon. In 1971, astronaut Alan Shepard hit two golf balls during the Apollo 14 mission.

652. The silhouette on the NBA logo is based on Hall of Fame player Jerry West.

653. Wilt Chamberlain, one of the greatest basketball players, never fouled out of a game during his entire career.

654. The longest tennis match took place at Wimbledon in 2010, lasting 11 hours and 5 minutes over three days. John Isner defeated Nicolas Mahut in this epic battle.

655. During World War II, the Pittsburgh Steelers and the Philadelphia Eagles combined to form a team called the "Steagles" due to player shortages.

656. Georgia, the Redneck Games are a celebration of quirky and humorous competitions. The event started in 1996 as a lighthearted response to the Summer Olympics held in Atlanta that year. Some of the unique events include Mud Pit Belly Flop: Contestants perform belly flops into a pit of mud, with the biggest splash winning—they also play toilet Seat Horseshoes.

657. Frozen Cow Poo Puck: The first hockey pucks used in early outdoor hockey games were made of frozen cow dung.

658. Black Underwear Rule: Major League Baseball umpires are required to wear black underwear in case they split their pants during a game.

659. Mike Tyson is an American boxer with an odd habit. Known for his ferocity in the ring, Tyson owned and cared for several pet tigers.

660. Created in 1895 by William G. Morgan in Holyoke, Massachusetts, volleyball was designed as a less strenuous alternative to basketball—that doesn't seem to have been the case.

661. Initially played by Native American tribes, lacrosse is considered the oldest sport in North America, with a history of over a millennium.

662. Emerging in the 1950s in California, skateboarding has grown from a niche activity to a globally recognized sport.

663. Chad Caruso set the record for crossing the US on a skateboard. He skated from Venice Beach, California, to Virginia Beach, Virginia, in just 57 days, 6 hours, and 56 minutes.

664. The record for the fastest speed on a skateboard was broken in 2017 when English skater Peter Connolly achieved 91.17 miles per hour on his board.

665. Colorado is the only state in history to turn down hosting the Olympics when offered in 1976.

22

JUST ODD

666. In early colonial New England, pumpkin shells were sometimes used as templates for haircuts to ensure a round and uniform finished cut. As a result of this practice, New Englanders were sometimes nicknamed "pumpkinheads."

667. Q is the odd one out when it comes to state names. It is the only letter not in one of the state's names.

668. The White House in Washington, D.C., has its own zip code (20500).

669. The General Electric building in Schenectady, New York, has its own zip code (12345). That would be easy to remember.

670. How do you like your hug? Tim Harris is the inspiring owner of Tim's Place in Albuquerque, New Mexico. Tim, who has Down syndrome, opened his restaurant in 2010 and quickly became known for his warm hugs and friendly atmosphere. He even keeps a hug counter on the wall to keep track of how many hugs he's given out. How's that for spreading the love?

671. The Atlantic City Boardwalk in New Jersey is considered the world's longest boardwalk, stretching over a whopping 5.5 miles!

672. In 1926, Ford Motor Company became one of the first companies in America to adopt a five-day, 40-hour workweek for its automotive labor and factory workers. Ford decided that his employees deserved a little R&R, so he gave them Saturdays and Sundays off.

673. Dave Thomas, the man behind the Wendy's empire, decided to hit the books again later in life. At the ripe old age of 61, he went to school and got his GED. Dave didn't want to be a dropout success story encouraging kids to drop out.

674. The Lincoln-Kennedy Paradox is a fascinating collection of coincidences between the lives of U.S. Presidents Abraham Lincoln and John F. Kennedy. Both were elected to Congress in '46 (Lincoln in 1846, Kennedy in 1946) and later to the presidency in '60 (Lincoln in 1860, Kennedy in 1960). Each president's last name has seven letters. Both were shot in the head on a Friday in the presence of their wives. Their successors were both named Johnson (Andrew Johnson for Lincoln, Lyndon B. Johnson for Kennedy), and both were born in '08.

675. Steven Wallace from the USA produced the loudest clap on record, measuring 117.4 decibels (dBA) on November 15, 2021, in Cambridge, Massachusetts.

676. Interestingly, formal wear and special occasion outfits tend to be kept the longest but are worn the least.

677. In 1956, Samuel Shenton founded the Flat Earth Society, which gained some notoriety in the 1960s during the Space Race. He was a frequent guest on television and in newspapers where he would cheerfully insist that the rest of the world was being duped by scientists.

678. The mustache cup! This ingenious invention from the Victorian era was designed to help men with mustaches enjoy their tea without getting their facial hair wet or ruining their carefully styled whiskers.

679. In 2014, Colorado officials changed the mile marker "420" on Interstate 70 to "419.99" to prevent it from being stolen repeatedly.

680. The record for the loudest dog bark in the United States is held by a Golden Retriever named Charlie on October 20, 2012. His bark reached an astonishing 113.1 decibels, about as loud as a jackhammer.

681. The loudest collective dog bark was recorded in Washington Park, Colorado, where 76 dogs barked together, reaching a volume of 124 decibels. That's louder than a rock concert!

682. The highest altitude achieved by a single kite in the United States is 14,121 feet above ground level. This record was set by a big Dunton-Taylor delta kite during a series of flights in September 2011.

683. The Kingda Ka at Six Flags Great Adventure in New Jersey is the tallest roller coaster in the world, with a height of 456 feet (139 meters). It also features a drop of 418 feet, making it one of the most thrilling rides.

684. When the founders of the company, Bill Hewlett and David Packard, started their business in 1939, they decided to name the company after themselves. Interestingly, they couldn't decide whose name should be first, so they settled the matter with a coin toss.

685. The oldest known blue jeans were discovered in a shipwreck off the coast of North Carolina. These jeans were found in the wreck of the SS Central America, which sank in 1857. The jeans, believed to be miner's work pants, feature a five-button fly and sold for $114,000.

686. Disneyland does not sell gum. Walt Disney implemented this policy to help keep the parks clean and ensure that guests don't have to deal with gum stuck to their shoes or other surfaces.

687. On October 14, 1912, while campaigning for the presidency in Milwaukee, Wisconsin, Theodore Roosevelt was shot in the chest by a would-be assassin named John Schrank. Despite being wounded, Roosevelt insisted on delivering his scheduled speech. He famously began by saying, "Friends, I shall ask you to be as quiet as possible. I don't know whether you fully understand that I have just been shot—but it takes more than that to kill a Bull Moose". He then spoke for 90 minutes before agreeing to go to the hospital.

688. Dog food tasters are real! Their job involves evaluating the taste, texture, and nutritional value of pet food to ensure it meets quality standards. Many dog food tasters have degrees in food science or nutrition.

689. An American Model plane 'RVJet Flying Wing', set a world record for the highest altitude of a remote-controlled (RC) model aircraft flight. In 2019, it reached an altitude of 34,800 feet at Spaceport America in New Mexico.

690. In 1987, Mr. T caused quite a stir in Lake Forest, Illinois, by cutting down over a hundred oak trees on his estate. This incident was famously dubbed the "Lake Forest Chain Saw Massacre" by the local media.

691. Professional sleepers are well-rested employees! Professional sleepers often participate in scientific studies and product Testing.

692. Aurora Ice Museum in Fairbanks, Alaska. This museum is the largest year-round ice environment in the world featuring mind-blowing carvings, including jousting knights, polar bear bedrooms, and even an ice bar where you can enjoy cocktails in icy martini glasses.

693. In New Jersey, In 1987, a psychic played a crucial role in helping to catch a murderer. A local psychic had horrific visions that led the police to Koedatich, who had killed two young women, Amie Hoffman and Deirdre O'Brien.

694. Casinos are designed to keep you engaged and playing for as long as possible, and one way they do this is by not having clocks or windows.

695. The sheep were removed from Central Park's Sheep Meadow in 1934. The city was concerned about their safety during the Great Depression, fearing they might be stolen for food.

696. Detachable collars on men's shirts have an interesting history! They were invented in 1827 by Hannah Montague in Troy, New York. She devised the idea to avoid washing her husband's entire shirt when only the collar was dirty.

697. Math professor Joan Ginther won four lotteries. And is often referred to as the "luckiest woman alive" due to her incredible streak of winning the lottery four times, amassing over $20 million in total. While some speculate that her mathematical expertise played a role in her wins, the exact methods she used remain a mystery.

698. In the 18th century, men originally wore high heels, which symbolized status and power.

699. Archaeologists discovered a 32-foot-long ship hull, along with the bow, stern, and anchor. An18th-century ship that was unearthed at the Ground Zero site in 2011!

The discovery was quite unexpected, as the ship was buried beneath what would become the parking garage for the new World Trade Center.

23

MEDICAL / HEALTH

700. This is not the way to fix OCD. A man named George was tormented by his OCD and attempted suicide with a .22-caliber rifle. The bullet lodged in his left frontal lobe. Remarkably, after the surgery to remove the bullet, George's OCD symptoms vanished without causing any other permanent brain damage. He went on to become a straight-A college student and lived a relatively normal life.

701. Hardened his heart. In a bizarre medical case in 2021, a 56-year-old man survived having cement accidentally injected into his heart during a spinal surgery. The cement had leaked into his veins and traveled to his heart, but he remarkably survived after emergency surgery.

702. Have you thought about catching HSAM or Highly Superior Autobiographical Memory? The first known case was Jill Price (in 2006). She can remember what she did, where she went, and even what she wore on any given day. Since then, more people have been identified with this extraordinary ability. The exact cause of HSAM remains a mystery.

703. During the Manhattan Project, 18 people were injected with plutonium as part of secret experiments conducted between 1945 and 1947. These experiments were carried out without the consent of the participants, who included men, women, and children

704. Loren Montefusco is a 22-year-old from South Carolina. Loren suffers from aquagenic urticaria, a rare condition where contact with water triggers a severe allergic reaction. Even the slightest touch of water causes her to experience a burning itch deep beneath her skin. She has to drink water through a straw to avoid water touching her skin.

705. Human Stomach Acid: The acid in your stomach is strong enough to dissolve razor blades. Its pH level is 1.0 to 2.0, making it incredibly potent.

706. No more ice for me. In 2006, a 12-year-old girl named Jasmine Roberts made a surprising discovery during her science project. She found ice served at fast food restaurants was often dirtier than toilet water.

707. The use of early syringes by Native Americans influenced modern medical technology. Their innovative use of hollow bird bones and animal bladders to inject fluids into the body laid the groundwork for the development of more advanced syringes.

708. Did we get aspirin from Native Americans? Native Americans used the bark of the willow tree as a pain reliever. The bark contains salicin, a natural compound that is chemically similar to aspirin. The discovery of salicin in willow bark eventually led to the development of aspirin, a widely used and effective pain reliever in modern medicine.

709. In 1952, it was and American, Dr. Virginia Apgar who Created the Apgar Score, a quick and simple way to assess the health of newborns immediately after birth.

710. African American Dr. Charles Drew's contribution to medicine in 1938 is still saving lives. He pioneered methods of storing blood plasma, which led to the development of large-scale blood banks.

711. Goldenseal and other plants were used by native Americans for their antiseptic properties to clean wounds.

712. In 1849 Dr. Elizabeth Blackwell earned her medical degree from Geneva Medical College in New York. She was the first woman in the United States to receive a medical degree, paving the way for future generations of women in medicine.

713. In 1864 Dr. Rebecca Lee Crumpler became the first African American woman to become a physician in the United States.

714. It is believed that the first successful open-heart surgery in America was performed by Dr. Daniel Hale Williams on July 9, 1893. Dr. Williams, a black surgeon, operated on a man named James Cornish, who had been stabbed in the chest. Without the aid of modern surgical tools or anesthesia, Dr. Williams successfully sutured Cornish's pericardium, the sac around the heart, making it one of the first recorded heart surgeries. Cornish went on to live for another 20 years.

715. In 1888, Dr. William Williams Keen performed the first successful brain surgery in America. Dr. Keen was a pioneering surgeon who successfully removed a brain tumor, marking a significant milestone in the field of neurosurgery.

716. Forest bathing, or shinrin-yoku, is a Japanese practice that involves immersing yourself in nature to soak up its health benefits. Forest bathing encourages mindfulness and being present in the moment. It involves walking slowly, breathing deeply, and engaging all your senses to connect with nature.

717. Jonah's Complex concept, introduced by psychologist Abraham Maslow, refers to the fear of success or the fear of realizing one's full potential.

718. Randy Gardner, a 17-year-old high school student from San Diego, California, completed the longest time a person has stayed awake (in America). In December 1963 and January 1964, he stayed awake for 11 days and 24 minutes (264.4 hours) as part of a science fair project.

719. On average, it takes most people between 10 to 20 minutes to fall asleep. if it's taking you much longer than 20 minutes, it could be a sign of an underlying sleep disorder or other issues affecting your sleep.

720. Dr. William Osler - Often called the "Father of Modern Medicine," he was one of the founding professors of Johns Hopkins Hospital.

721. In 1888, Dr. William Williams Keen performed the first successful brain surgery in America. Dr. Keen was a pioneering surgeon who successfully removed a brain tumor, marking a significant milestone in the field of neurosurgery.

722. Leeches administered by your doctor. The FDA approved the use of leeches for medical purposes in 2004. Medicinalis leeches, specifically the species Hirudo Medicinalis, are used primarily to help with localized venous congestion after surgery.

723. MAGGOTS? The FDA approved the use of maggots for medical purposes in 2004. These maggots, specifically the larvae of the green bottle fly, are used to clean non-healing wounds by consuming dead tissue, which helps promote healing.

724. Your heart beats about 100,000 times a day, pumping around 2,000 gallons (7,570 liters) of blood through your body.

725. Laughter indeed has some remarkable health benefits, including potential effects on cancer cells. Research has shown that laughter can boost the activity of natural killer (NK) cells, which are a type of immune cell that plays a significant role in fighting cancer. By increasing the activity of these cells, laughter can help the body combat tumor cells more effectively.

726. In 1972 William C. Dement and his colleagues at Stanford University created the Sleepiness Scale (SSS) a tool designed to measure levels of sleepiness throughout the day.

727. Ayurveda: This ancient Indian system of medicine focuses on balancing the body's energies (doshas) through diet, herbal treatments, and lifestyle changes. It's becoming more popular for its holistic approach to health.

728. Sound Healing: This practice uses sound vibrations from instruments like singing bowls, gongs, and tuning forks to promote relaxation and healing. It's rooted in ancient traditions from various Eastern cultures.

729. The primary cause of microsleep is sleep deprivation. Even a single night of restricted sleep can increase the likelihood of experiencing microsleep.

730. During microsleep, brain activity changes, with some parts of the brain entering a sleep state while others remain awake. This can lead to impaired cognitive function and reaction times.

731. Broken heart syndrome, also known as takotsubo cardiomyopathy or stress-induced cardiomyopathy, is a real medical condition. It occurs when extreme emotional or physical stress leads to a sudden, temporary weakening of the heart muscle. It often mimics a heart attack, with symptoms like chest pain and shortness of breath.

732. Cherries are not just delicious; they might also be helping us fight the big C. According to some studies, cherries are packed with all sorts of goodies like melatonin, ursolic acid, and anthocyanins, which have been shown to inhibit the growth and proliferation of breast cancer cells.

733. Here's an intriguing medical phenomenon: Auto-Brewery Syndrome. This rare condition causes a person's gut to produce alcohol from the carbohydrates they consume. Essentially, their digestive system acts like a mini-brewery, leading to symptoms of intoxication without drinking any alcohol.

734. Remember Dr. Jonas Salk - Developed the first successful polio vaccine in 1954, which has saved countless lives worldwide.

735. Here is an odd and rare one: Foreign Accent Syndrome. causes individuals to suddenly start speaking with a foreign accent, often after a head injury, stroke, or other neurological event. The accent is usually not a perfect imitation, but it can be quite distinct and persistent.

736. Your bones are constantly regenerating. Every ten years or so, you essentially get a new skeleton because your bones are in a constant state of renewal. This process, known as bone remodeling, involves the breakdown of old bone tissue and the formation of new bone tissue.

737. One of the newest and most exciting medical breakthroughs in 2024 is the development of a new Alzheimer's drug by Eli Lilly. In clinical trials, it has been shown to slow cognitive decline by 35% and reduce the decline in daily activities by 40%.

738. Another exciting recent medical discovery is the development of personalized cancer vaccines. These vaccines are designed to prime the immune system to target cancer cells specific to an individual's tumor.

739. Gene Therapy for Blood Disorders: Following the success of CRISPR-based treatments for sickle cell anemia, similar therapies are being developed for other genetic blood disorders, such as beta-thalassemia.

740. Neuralink has begun implanting its brain chip in human patients, which shows promise to help with several neurological challenges.

741. First Successful Organ Transplant: In 1954, Dr. Joseph Murray performed the first successful human kidney transplant in Boston, Massachusetts. This pioneering surgery opened the door to modern organ transplantation.

742. CRISPR Gene Editing: In 2012, Jennifer Doudna and Emmanuelle Charpentier, working at the University of California, Berkeley, developed the CRISPR-Cas9 gene-editing technology.

743. According to some studies, the hydrogen sulfide in farts, yes, the same gas that gives them that delightful aroma, might help regulate blood pressure. It's like your body's way of saying, "Hey, let's keep things chill down there."

744. AI is being used to tailor treatments to individual patients based on their unique genetic makeup and medical history. We need to monitor   AI.

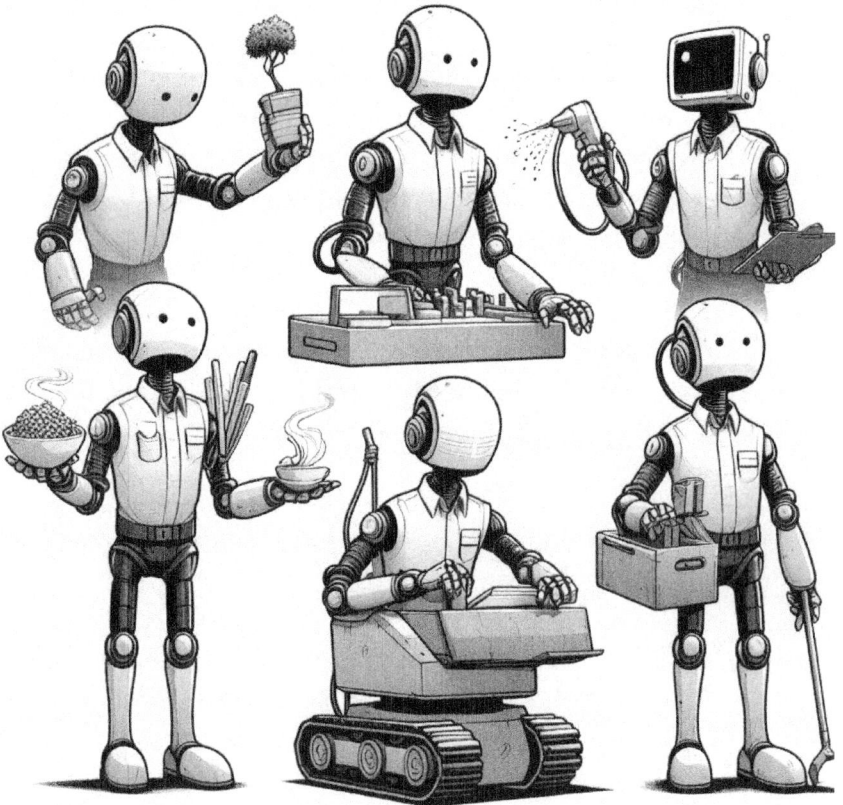

745. A patient in his 50s arrived for a routine CT scan when the Artificial intelligence-based program warned he might have intracranial bleeding. AI is saving lives.

24

INNOVATIONS & INVENTORS

746. The first record of running water being used in the United States was in the White House on its main floor in 1833.

747. Though invented in China, some flush toilets were reported to have been in use in NY as early as the 1820s.

748. The first (more modern) flush toilet in the United States was reportedly installed during Millard Fillmore's presidency in 1853.

749. Boston is credited with installing the first comprehensive sewer system in the United States in 1823. This early system was designed to carry stormwater and sewage away from the city streets to reduce the risk of disease. The initial sewer lines were constructed using brick and stone.

750. The first water treatment plant in the United States was built in Poughkeepsie, New York, in 1872. This plant used a slow sand filtration system to treat water, which was a significant advancement in public health and sanitation at the time.

751. While on this subject the first pay toilet in the United States was installed in 1910 in Terre Haute, Indiana.

752. Mary Kies was the first woman to receive a U.S. patent; on May 5, 1809, she secured a patent for a process of weaving straw with silk or thread, which was a game-changer for the hat and bonnet industry.

753. In Delaware, there are about 2 million registered corporations, with a state's population of only around 1 million.

754. The White House got electricity in 1891 during President Benjamin Harrison's administration. However, President Harrison and his wife Caroline were so afraid of getting an electric shock that they refused to touch the light switches themselves. The White House staff was in charge of turning the lights on and off.

755. The first electric refrigerator was installed at the White House during the Calvin Coolidge administration (1926). Be Cool with Coolidge.

756. The White House switched from direct current (DC) to alternating current (AC) during a major renovation that took place between 1948 and 1952 under President Harry S. Truman.

757. The first message sent on the telegraph was "What hath God wrought?" Samuel Morse transmitted this historic message from Washington, D.C., to Baltimore on May 24, 1844.

758. In 1876, the first-year phones were available, the cost of a 3-minute phone call was around $20.70. Adjusted for inflation, that's a whopping $500 in today's money.

759. The first phone call was made by Alexander Graham Bell on March 10, 1876. He made this historic call from his laboratory in Boston, Massachusetts, to his assistant, Thomas Watson, who was in another room. Bell's famous words were, "Mr. Watson, come here, I want to see you."

760. In 1956, a guy named Malcolm McLean had a brilliant idea. He thought, "Hey, what if we just put all the stuff we want to ship in big metal boxes and stack them on the ship?" And thus, the **shipping container** was born! Suddenly, ships could carry more cargo, and it was easier and cheaper to move stuff around the world. They could be loaded onto trucks, trains, and ships.

761. Henry F. Phillips didn't invent the screw and screwdriver design himself! The original design was patented by a guy named John P. Thompson in 1932, but he couldn't find any manufacturers interested in his creation. So, he sold the rights to Phillips, who then refined the design and made it popular. The self-centering property of the Phillips design made it easier for power tools to drive the screws, speeding up the production process.

762. The term "duck tape" itself, referring to strips of plain cotton duck cloth, was in use during the 1800s. This fabric was known for its durability and water-resistant properties, which made it useful for various applications, including wrapping steel cables and reinforcing clothing.

763. In the late 1800s, Johnson & Johnson was already making significant strides in the field of adhesive products. While the modern version of duct tape wasn't invented until World War II, the company was producing adhesive plasters and tapes as early as the late 19th century.

764. Baby cages? also known as "health cages," were invented in the United States in 1922 by Emma Read. The cages were wire beds that could be attached to the outside of a building, allowing parents to place their babies inside to get fresh air and sunshine.

765. The Yodel Meter, a fascinating gadget from the 1920s, was designed to measure the pitch of the human voice, particularly useful for those who wanted to perfect their yodeling skills. Imagine a time when people were so passionate about yodeling that they needed a specialized device to ensure their yodels were pitch-perfect!

766. Now, this is important. Spaghetti Aid, a quirky invention from the 1950s, was designed to simplify the process of eating spaghetti. While it aimed to make spaghetti consumption easier, its effectiveness in achieving this goal remains questionable.

767. Only in Texas!! James Alexander Williams of Texas patented the gun-powered mousetrap in 1882. It was a curious invention that incorporated a handgun.

768. The Dog-Restraining Device, a mechanical extendable arm designed to restrain dogs, was an invention from the 1940s. Despite its creation, it was largely unnecessary because Mary Delaney had already patented the more practical idea of a dog leash in 1908.

769. 1857, Joseph C. Gayetty introduced a product called "Medicated Paper, for the Water-Closet," now called toilet paper.

770. It sounds like something they would come up with in California. Slugbot—an AI robot designed to crawl over soil and pick up slugs, using them as an organic power source.

771. The first moving sidewalk was introduced at the World's Columbian Exposition in Chicago, Illinois, in 1893.

772. The Selfie Toaster was a quirky invention that allowed you to upload a photo of yourself. The photo would then be etched onto a metal plate and inserted into the toaster. The result? A piece of toast with your mug on it.

773. Beachwear, made from tanned and dyed salmon skin, was introduced in 2003.

774. A pogo stick that uses an air pump instead of springs was introduced in 2001.

775. The Ecopod was designed to decompose naturally, providing a final resting place that was as eco-friendly as it was stylish. It even came in a variety of colors so that you could match your coffin to your favorite outfit.

776. In 1948, the first atomic clock was developed at the United States's National Bureau of Standards.

777. King Camp Gillette invented disposable razors in 1904. It is estimated that over 2 billion disposable razors are disposed of every year in the United States.

778. The first female American millionaire was Madam C. J. Walker (born Sarah Breedlove; December 23, 1867 – May 25, 1919). She made her fortune by developing and marketing a line of cosmetics and hair care products for Black women.

779. That's being a little on the lazy side. A cone that rotates the ice cream for you, so you don't have to lick around it. (1998)

780. The innovative invention by Cassidy Matwiyoff, a high school senior from San Diego, California. Cassidy introduced her invention, the Bowwow, a robot designed to pet an owner's furry friend while they're out of the house. (1923)

781. The Swatch watch, introduced in 1983, revolutionized the watch industry with its unique, fun design and affordable price.

782. Vulcanized Rubber: Charles Goodyear's discovery of the vulcanization process in 1839 made rubber more durable and elastic, leading to its widespread use in various industries.

783. Walter Hunt invented the safety pin in 1849, a simple yet ingenious device that has remained in use ever since.

784. Elias Howe is credited with inventing the modern lockstitch sewing machine in 1846. His machine had a needle with the eye at the point, a shuttle operating beneath the cloth to form the lock stitch and an automatic feed.

785. A Brookings Institution report states that 47% of U.S. workers will likely see their jobs automated over the next 20 years.

Optimus is a humanoid robot being developed to take over all the jobs that nobody wants to do. Goodbye, minimum-wage workers; hello, robot overlords! Let's get together in 2025 and see what Optimus is doing.

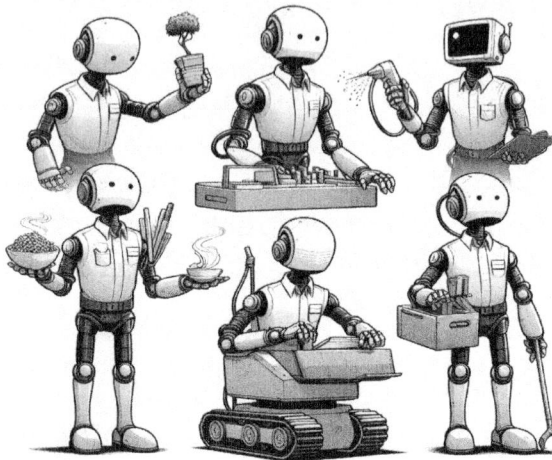

25

SCIENCE

786. The first dinosaur fossil found in the United States was a Megalosaurus, discovered in 1824 in the state of Massachusetts. Megalosaurus, meaning "great lizard," was a large meat-eating dinosaur that lived during the Jurassic period. The discovery of this dinosaur fossil was a significant event in the history of paleontology, as it provided evidence that dinosaurs had once roamed the Earth.

787. The Global Positioning System (GPS) is operated and maintained by the U.S. Air Force. It provides crucial positioning, navigation, and timing services worldwide.

788. Between the 1940s and 1970s, The University of California, Berkeley, discovered 16 elements of the periodic table.

789. The MRI machine was invented in 1977 by Dr. Raymond V. Damadian. He was a physician and scientist at the Downstate Medical Center, State University of New York in Brooklyn.

790. The United States has produced the most Nobel Prize winners in science, with numerous laureates in fields like physics, chemistry, and medicine.

791. Man of steel. The femur, or thigh bone, is the largest and strongest bone in the human body. It can support up to 30 times the weight of your body, making it stronger than steel ounce for ounce.

792. The discovery that launched modern paleontology was the finding of the first nearly complete dinosaur skeleton, Hadrosaurus foulkii, in 1858 in Haddonfield, New Jersey.

793. Your speeding! Messages from the human brain travel along nerves at up to 200 miles per hour.

794. The first woman to earn a PhD in Computer Science in the US was Sister Mary Kenneth Keller. She received her doctorate from the University of Wisconsin-Madison in 1965.

795. In 2014, Illinois became the first U.S. state to ban the manufacture and sale of cosmetics containing plastic microbeads. These tiny plastic beads, often found in exfoliating face washes, are harmful to the environment because they can pass through water treatment systems and end up in rivers, lakes, and oceans.

796. Tesla Coil Although not used directly in everyday applications, the Tesla coil is a fundamental component in various technologies, including radio transmission and medical devices like X-ray machines.

797. Tesla's pioneering experiments with wireless transmission of electricity and radio waves demonstrated the principles of wireless communication, which are the basis for modern technologies like Wi-Fi and Bluetooth.

798. Tesla developed the AC electrical system, which is the standard for power transmission and distribution worldwide.

799. The first commercial nuclear power plant in the United States was the Shippingport Atomic Power Station. President Dwight D. Eisenhower opened it on May 26, 1958, as part of his Atoms for Peace program.

800. Bluetooth technology was invented in 1994 by a team of engineers at Ericsson, led by Dr. Jaap Haartsen. The first Bluetooth-enabled device, a hands-free mobile headset, was introduced in 1999.

801. Vint Cerf and Robert Kahn: They are often credited with inventing the fundamental communication protocols, TCP/IP (Transmission Control Protocol/Internet Protocol), which allow different networks to connect and communicate. Their work laid the foundation for the modern internet.

802. The concept of the internet was first developed in the United States. ARPANET, the precursor to the internet, was created by the U.S. Department of Defense in the late 1960s.

803. Tim Berners-Lee: He invented the World Wide Web in 1989, a system of interlinked hypertext documents accessed via the Internet.

804. On April 3, 1973, Martin Cooper, an engineer at Motorola, invented the first portable cell phone and made the world's first cell phone call.

805. The Crystal Geyser is located near the Green River, Utah. is a **cold-water geyser** driven by carbon dioxide gas. It erupts due to the buildup of carbon dioxide gas in the underground water. Interestingly, the Crystal Geyser was accidentally created during oil exploration in the 1930s. Drillers hit an aquifer, and the resulting pressure created the geyser.

806. Forensic ballistics is the field that applies ballistics principles to solve crimes. The St. Valentine's Day Massacre in 1929 highlighted the need for forensic ballistics to match bullets to specific firearms.

807. 1902: Dr. Henry P. de Forest of the New York Civil Service Commission started using fingerprints to identify applicants for civil service jobs.

808. Professor Edwin B. Frost and Dr. Gilman D. Frost performed the first clinical X-ray in the United States on January 20, 1896, at Dartmouth College.

809. The metal Gallium has a melting point of around 85.5°F (29.7°C), which is just below our body temperature. So, if you hold it in your hand for long enough, it will indeed melt like an M&M in your mouth. But unlike an M&M, you can't eat gallium, no matter how tempting it might be.

810. In 2015, scientists at the Laser Interferometer Gravitational-Wave Observatory (LIGO) in the United States detected gravitational waves for the first time. This confirmed a major prediction of Albert Einstein's general theory of relativity.

811. Completed in 2003, this international research project, led by the United States, successfully mapped the entire human genome.

812. The first solar-powered satellite was Vanguard, launched by the United States on March 17, 1958. This small satellite, about the size of a grapefruit, was designed to test the capabilities of a three-stage launch vehicle and the effects of the environment on a satellite and its systems in Earth orbit.

813. Humans Only Use 10% of Their Brains: The myth that humans only use 10% of their brains has been debunked by neurological research showing that virtually all parts of the brain have known functions.

814. One of the largest dinosaurs discovered in the United States is the Diplodocus carnegii, often referred to as "Dippy." This massive dinosaur was unearthed in Albany County, Wyoming, in 1899. The skeleton measures about 87 feet long and was one of the first nearly complete dinosaur skeletons ever found.

815. The Chicago Pile-1: In 1942, the world's first artificial nuclear reactor, Chicago Pile-1, was built under the stands of Stagg Field at the University of Chicago. It marked the beginning of the atomic age.

816. They are amongst us. CaliBurger in Pasadena, California, features the Flippy robot, developed by Miso Robotics. Flippy handles grilling and frying tasks, and it can cook burgers, fries, and other items with precision and consistency.

817. Xcelsior AV, developed by New Flyer in collaboration with Robotic Research. This autonomous (without a human driver) electric transit bus is being tested in a pilot project with the Connecticut Department of Transportation.

818. Tipsy Robot in Las Vegas, Nevada, offers a unique experience with robotic bartenders who mix and serve drinks with unmatched precision.

26

SPACE EXPLORATION

---

819. Maria Mitchell (1818-1889): The first professional female astronomer in the United States, she discovered a comet in 1847, named "Miss Mitchell's Comet."

820. Asaph Hall was an American astronomer who, in 1877, discovered the two moons of Mars, Deimos and Phobos.

821. Edward Emerson Barnard (1857-1923): Famous for his discovery of Barnard's Star and his extensive work on dark nebulae (the birthplace of stars).

822. The only two planets in our solar system that do not have moons are Mercury and Venus.

823. Earth's Rotation: Earth's rotation is gradually slowing down. This means that days are getting longer, but it happens at such a slow rate that it will take about 140 million years for the length of a day to increase by one hour.

824. Earth is about 18 galactic years old. A galactic year is the time it takes for the Milky Way to rotate around its center, which is approximately 230 million Earth years.

825. The state of New Mexico passed a resolution declaring Pluto a planet whenever it passes overhead in their night skies.

826. Discovery of Pluto: In 1930, Clyde Tombaugh discovered Pluto at the Lowell Observatory in Flagstaff, Arizona.

827. Launched in 1990, the Hubble Space Telescope has provided stunning images and invaluable data about the universe, leading to numerous discoveries about galaxies, black holes, and the expansion of the universe.

828. Henry Draper, a pioneer in astrophotography, took the first photographs of the Orion Nebula and the spectrum of stars.

829. In 2019, the Event Horizon Telescope collaboration, which included significant contributions from American scientists, captured the first-ever image of a black hole. This historic achievement provided direct visual evidence of these mysterious objects.

830. Ormsby MacKnight Mitchel (1809-1862): Known as the "Carl Sagan of the 1800s," he was a dynamic speaker and popularizer of astronomy.

831. William Cranch Bond (1789-1859) was the first director of the Harvard College Observatory to discover Hyperion, a moon of Saturn, and was a pioneer in the use of photography in astronomy.

832. Established in 1897, it was the first observatory to be built, with the primary function being research as opposed to merely housing a telescope.

833. The Astrophysical Journal, often abbreviated as ApJ, is a peer-reviewed scientific journal that has been a cornerstone of astronomy and astrophysics since its founding in 1895.

834. Stellar spectroscopy (1800s) The light we see from stars is actually a mix of different colors, each coming from different elements in the star's atmosphere. By looking at how much of each color is present, astronomers can figure out what the star is made of.

835. The Moon is slowly drifting away from Earth: Each year, the Moon moves about 3.8 centimeters (1.5 inches) farther from our planet.

836. The Moon experiences moonquakes. These are caused by tidal forces from Earth and can last up to an hour.

837. The Moon has a smell: Astronauts who walked on the Moon reported that lunar dust smelled like burnt gunpowder. HOW

838. The Moon is the fifth largest moon in the Solar System. Despite being relatively small compared to Earth, it is quite large compared to other moons in our Solar System.

839. Mars has the largest volcano in the solar system: Olympus Mons is about 13.6 miles (22 kilometers) high, nearly three times the height of Mount Everest.

840. In 2013, the Curiosity Rover celebrated its Martian birthday by playing "Happy Birthday" to itself using its Sample Analysis at Mars (SAM) instrument. It's like the rover was saying,

841. Playtex, the company known for making bras and girdles, was responsible for designing the Apollo spacesuits. They had to be both flexible and durable[1].

842. Due to its axial tilt, Mars has four seasons like Earth. However, each season lasts about twice as long because a Martian year is nearly twice as long as an Earth year.

843. Due to the fine dust in its atmosphere, sunsets on Mars appear blue to the human eye.

844. The crew of Apollo 7 received an Emmy for their live television broadcasts from space, making them the only astronauts to win this prestigious TV award.

845. During the Apollo missions, astronauts loved to eat bacon squares, specially prepared bacon cubes that were easy to eat in zero gravity.

846. Alan Shepard became the first American to travel into space on May 5, 1961, aboard the Freedom 7 spacecraft.

847. John Glenn orbited Earth thrice on February 20, 1962, in the Friendship 7 spacecraft.

848. On July 20, 1969, Neil Armstrong and Buzz Aldrin became the first humans to walk on the Moon during NASA's Apollo 11 mission.

849. On June 18, 1983, Sally Ride flew aboard the Space Shuttle Challenger, becoming the first American woman in space.

850. During the Apollo 14 mission, astronaut Stuart Roosa took hundreds of tree seeds to the Moon. These seeds were later planted on Earth and grew into "moon trees".

851. The first private spacecraft to dock with the ISS: SpaceX's Dragon spacecraft, launched by an American company, became the first commercial spacecraft to dock with the ISS in 2012.

852. From 1981 to 2011, NASA's Space Shuttle program conducted 135 missions, including constructing the International Space Station (ISS).

853. NASA's "vomit comet" To simulate zero gravity, NASA uses a plane that flies in parabolic arcs, creating brief periods of weightlessness. This plane is affectionately known as the "vomit comet" because it often makes passengers nauseous.

854. Buzz Aldrin took the first space selfie during the Gemini 12 mission 1966. It's a classic shot of him with Earth in the background.

855. Despite its popularity on Earth, freeze-dried "astronaut ice cream" was never actually eaten by astronauts in space.

856. It's often said that NASA spent millions developing a pen that could write in space, while the Soviets just used pencils. In reality, both space agencies used pencils initially but later adopted the Fisher Space Pen,

857. Shortly after launch, Apollo 12 was struck by lightning twice. The quick thinking of flight controller John Aaron, who suggested "SCE to AUX," saved the mission. This obscure command reset the spacecraft's electrical system.

858. The ISS orbits the Earth at a blistering speed of about 17,100 miles per hour, which means it completes one full orbit every 90 minutes.

859. Apollo 16 astronaut Charles Duke left a framed family photo on the Moon's surface. The back of the photo reads, "This is the family of astronaut Charlie Duke from planet Earth, who landed on the moon on April 20, 1972." Exposure to the harsh lunar environment has likely bleached the photo white by now.

860. During the Apollo 10 mission, astronauts reported hearing strange "outer space-type" music while orbiting the far side of the Moon. This eerie sound was later attributed to radio interference, but it spooked the astronauts.

861. The Toss Zone: After completing their tasks on the Moon, Apollo 11 astronauts Neil Armstrong and Buzz Aldrin created a "toss zone" by discarding items they no longer needed, such as the TV camera and tools used to gather moon rocks. This was done to lighten the load for their return journey.

862. Launched in 1990, the Hubble Space Telescope has provided some of the most detailed images of space ever captured, revolutionizing our understanding of the universe.

863. Launched in 1977, Voyager 1 and 2 are NASA's longest-operating space missions and have traveled farther than any other human-made objects.

864. In 2001, Dennis Tito, an American engineer and entrepreneur, became the world's first space tourist, traveling to the ISS aboard a Russian Soyuz spacecraft.

865. As of recent surveys, about 5-10% of Americans believe that the moon landing was faked.

27

WEATHER

866. The "Big Blow" of 1921, also known as the Great Olympic Blowdown, was a hurricane-force windstorm that struck the coast of Washington on January 29, 1921. The storm's winds reached incredible speeds, with gusts estimated at 150 miles per hour (240 km/h) before the anemometer was blown away.

867. The Waffle House Index! The US government does use this as an unofficial measure of storm severity. If a Waffle House is open and serving a full menu, that's a green light. If it's open but with a limited menu, that's a yellow light. If a Waffle House is closed, that's a red light, meaning things are pretty dire.

868. The Heppner flood of 1903 was a devastating flash flood that struck Heppner, Oregon, on June 14, 1903. This tragic event remains the deadliest natural disaster in Oregon's history, with a death toll of 247 people. The flood was triggered by a severe thunderstorm that caused torrential rain and hail to fall on the watersheds of Willow Creek and its tributaries, leading to a massive wall of water rushing down the streams. Within minutes, a 15-to-50-foot wall of water swept through Heppner, destroying homes, businesses, and infrastructure.

869. During the Carr Fire near Redding (2018), California, a fire tornado with winds reaching 143 mph formed. This phenomenon, also known as a "firenado," was as intense as an EF3 tornado.

870. Donut-Shaped Hail in Wisconsin (2021): A hailstone with a perfect donut shape was found in Verona, Wisconsin. This unusual formation puzzled meteorologists and was a rare sight.

871. In 1995, large hailstones injured over 400 people during the Mayfest festival in Fort Worth, Texas.

872. Snow in Hawaii (2021): While snow in Hawaii isn't unheard of, the amount that fell in December 2021 was extraordinary. Mauna Kea and Mauna Loa received significant snowfall, creating a winter wonderland in the tropics,

873. The Great Blue Norther (1911): On November 11, 1911, a cold front swept through the central United States, causing temperatures to drop dramatically within hours. The temperature in Kansas City, Missouri, fell from 76°F to 11°F in just 10 hours.

874. The Tri-State Tornado (1925) This tornado, which traveled through Missouri, Illinois, and Indiana, is the deadliest in U.S. history. It covered a distance of 219 miles and caused 695 fatalities.

875. The Year Without a Summer (1816) Following the eruption of Mount Tambora in Indonesia, 1816 saw severe climate abnormalities. Snow fell in June in New England, and frosts occurred throughout the summer, leading to crop failures and food shortages.

876. The largest hailstone ever recorded fell near Vivian, South Dakota, on July 23, 2010. This massive hailstone measured 8 inches (20.3 cm) in diameter and weighed 1.94 pounds (0.88 kg).

877. The Enigma Tornado Outbreak of 1884 is one of U.S. history's most mysterious and devastating tornado events. On February 19-20, 1884, this outbreak swept across the Southeastern United States, producing an estimated 60 or more tornadoes. The tornadoes affected multiple states, including Alabama, Georgia, Illinois, Indiana, Kentucky, Mississippi, North Carolina, South Carolina, Tennessee, and Virginia. The death toll is estimated to be between 800 and 1,200 people, making it one of the deadliest tornado outbreaks in history.

878. The 1974 Super Outbreak occurred on April 3-4, 1974. The outbreak produced 148 tornadoes across 13 U.S. states and Ontario, Canada. It was the first outbreak to generate over 100 tornadoes in 24 hours. Among these tornadoes, 30 were classified as violent (F4 or F5), making it the most violent tornado outbreak ever recorded.

879. The 2011 Super Outbreak occurred from April 25 to 28, 2011. The outbreak produced 360 confirmed tornadoes across 21 states, making it the largest tornado outbreak ever recorded. Four of these tornadoes were rated EF5, the highest rating on the Enhanced Fujita scale, with winds exceeding 200 mph. The outbreak resulted in 321 fatalities and caused approximately $12 billion in damages (2021 dollars).

880. The United States experiences more tornadoes than any other country in the world. On average, the U.S. sees about 1,150 to 1,200 tornadoes annually

881. The "Human Lightning Rod," Roy Sullivan was a park ranger in Virginia who was struck by lightning **seven times** between 1942 and 1977. He survived all the strikes but suffered various injuries, including burns and the loss of a toenail.

882. People struck by lightning often develop a unique pattern on their skin known as Lichtenberg figures. These are branching, tree-like patterns created by the passage of high-voltage electrical discharges along the skin.

883. One of the earliest American meteorologists was **James Pollard Espy**. Often referred to as the "Storm King. He was the first official meteorologist appointed by the U.S. government, working for the War (1842) and Navy (1848) departments.

884. The record for the most snowfall in a 24-hour period in the United States is held by Silver Lake, Colorado. On April 14-15, 1921, an astonishing 75.8 inches (192.5 cm) of snow fell.

885. The record for the most rainfall in a 24-hour period in the United States was set in Alvin, Texas. On July 25-26, 1979, 42 inches of rain fell during Tropical Storm Claudette.

886. The record for the most rainfall in one hour in the United States was set in Holt, Missouri. On June 22, 1947, an astonishing 12 inches (304.8 mm of rain fell in just 42 minutes.

887. The Farmer's Almanac was first published in 1818. It was founded by David Young, a poet, astronomer, and teacher, along with publisher Jacob Mann. Since then, it has become a beloved publication, offering long-range weather predictions, gardening tips, humor, and fun facts for over two centuries.

888. Mount Rainier, Washington, holds the record for the most consecutive days with measurable snowfall in the United States. From February 13 to February 19, 1959, Mount Rainier experienced 6 consecutive days of snowfall.

889. Mount Rainier, Washington, holds the record for the most consecutive days with measurable snowfall in the United States. From February 13 to February 19, 1959, Mount Rainier experienced 6 consecutive days of snowfall.

890. February 13, 1958: Several counties in the Florida Panhandle, including Gadsden and Wakulla, recorded up to 3 inches of snow.

891. The hottest temperature ever recorded in Fairbanks, Alaska, was 96°F (35.6°C) on June 15, 1969. It's quite surprising to think of such high temperatures in a place known for its cold winters!

892. In the United States, an average of 25 million metric tons of road salt is used each year to melt ice and snow on roads. This amount is equivalent to nearly 28 million tons.

893. Some areas, like Polk County in Wisconsin, use cheese brine from local dairy companies. This salty byproduct from cheese production is sprayed on roads to prevent ice formation.

894. Developed during World War II, weather radar technology was initially used to detect enemy aircraft. However, scientists quickly realized its potential for meteorology. By the 1950s, weather radar was being used to detect precipitation, track storms, and provide early warnings for severe weather.

895. And, of course, AI and machine learning algorithms analyze large datasets to identify patterns and improve the accuracy of weather forecasts.

28

FAMILY

896. Having both red hair and blue eyes is extremely rare. Only about 0.17% of the world's population has this combination.

897. Helen and Les Brown, the couple who were born on the same day and had a love story that could rival any Hollywood romance. They were born on December 31, 1918, and were married for an impressive 75 years. died one day apart in July 2013. Helen passed away on July 16, and Les died the following day on July 17.

898. Barbara Soper from Michigan has an incredible story! She gave birth to three children in three consecutive years with very memorable dates:

- Chloe was born on 8/8/08.

- Cameron** arrived on 9/9/09.

- Cearra** was born on 10/10/10.

It makes it easy to remember birthdays.

899. The Coble family experienced a heartbreaking tragedy in 2007 when Lori and Chris Coble lost their three young children—Kyle, Emma, and Katie—in a car accident. (Two girls, one boy.) In 2008, they welcomed triplets—Ashley, Ellie, and Jake—into their lives (two boys and one girl). The Cobles believe their lost children are watching over their new siblings, bringing comfort and continuity to their family.

900. Legally, the offspring of identical twins are considered first cousins. However, genetically, they are more similar to full siblings.

901. As of 2023, the average number of children under 18 per family in the United States is 1.94. This is a decrease from the average of 2.33 children per family in 1960.

902. She had 14 babies in total. The record for the most children born to a single woman in the United States is held by Nadya Suleman, known as "Octomom." She gave birth to octuplets (eight babies) in January 2009, adding to six other siblings.

903. The heaviest baby ever recorded was born to Anna Bates in 1879. The baby weighed a whopping 22 pounds (10 kg) and measured 28 inches (71 cm) long.

904. . The Blair Family: Henry Blair, the second African American to hold a U.S. patent, invented a corn planter in 1834 that significantly increased farming efficiency. His contributions helped boost productivity for many farmers.

905. The 1st Ocen to Ocen family trip. In 1908, Jacob Murdock embarked on a remarkable journey with his family (and a mechanic) in a Packard "Thirty" touring car. The 7 set out from Los Angeles on April 24th and arrived in New York City on May 26th, covering 3,693 miles in 32 days.

906. Augusta Bunge and her family hold the record for the most generations in one picture, with **seven generations**[1]. This remarkable photo was taken in 1989 and includes Augusta, who was 109 years old at the time, along with her daughter, granddaughter, great-granddaughter, great-great-granddaughter, great-great-great-granddaughter, and great-great-great-great-grandson.

907. The title of the oldest family farm in the United States is often attributed to Tendercrop Farm (formerly Tuttle Farm) in Dover, New Hampshire. Established in 1632, it was continuously operated by the Tuttle family for 11 generations until it was sold in 20101.

908. In the United States, the Lilly Family Reunion in West Virginia is known for being one of the largest and longest-running family reunions. It has been held annually since 1929 and was recognized by the Guinness Book of World Records in 20092.

909. The Pioneering Smiths have a rich history of resilience and innovation. From their roots tracing the Mayflower, they have faced and overcome numerous challenges throughout history. They developed new farming techniques, built homes from local materials, and established trade networks with Native Americans. The Smiths have continued to contributing to various fields such as science, technology, business, and the arts.

910. The Chang family, Chinese immigrants who arrived during the California Gold Rush, transitioned from working on the railroads to becoming tech entrepreneurs in Silicon Valley.

911. The Patel family, who immigrated from India in the 1970s, started with a single convenience store and expanded into a thriving chain of businesses. Their success story is a classic example of the American Dream achieved through hard work and determination.

912. The Thompson family has produced numerous renowned musicians, from jazz legends in the Harlem Renaissance to contemporary pop stars. Richard Thompson is known for his distinctive guitar-playing style and his songwriting. Teddy, the son of Richard Thompson, is also a singer-songwriter. Kami Thompson - The daughter of Richard Thompson, Kami is also a singer-songwriter.

913. The Parkers, who own a large ranch in Texas, shifted from traditional cattle farming to leading in sustainable and organic farming practices.

914. The Johnson family, African American educators, have been instrumental in founding schools and colleges in the South, providing opportunities for countless students to achieve higher education and break the cycle of poverty.

915. George Washington Carver, born into slavery, became one of the most prominent agricultural scientists in American history. His work on crop rotation and soil improvement helped many Black sharecroppers diversify their crops and improve their livelihoods.

916. Robert Lloyd Smith, born during slavery, founded the Farmers' Home Improvement Society in 1890. This cooperative helped Black farmers improve their farming practices and build wealth through savings and efficient farming techniques.

917. The Adventurous O'Neills are a family of Irish descent with a history of adventure and exploration, from pioneering the Oregon Trail to modern-day travel blogging and adventure sports.

918. The Hernandez family, originating in Puerto Rico, have made significant contributions to science and technology. From pioneering medical research to developing cutting-edge tech, their story is one of innovation and excellence.

919. One of America's most prominent known families is the **Duggar family** from Arkansas. Jim Bob and Michelle Duggar have 19 children, all born between 1988 and 2009[1]. Their family gained widespread attention through the reality TV show "19 Kids and Counting."

920. The Bates family is from Tennessee and has 19 children as well. Gil and Kelly Bates, like the Duggars, have also been featured in their reality TV series, "Bringing Up Bates."

921. Thomas Mellon, the patriarch, was born in 1813 in Northern Ireland. He was the first in his family to graduate from college and went on to become a lawyer. He then founded T. Mellon & Sons' Bank in 1870, eventually becoming Mellon National Bank and Trust Company. The Mellons were involved in various industries, including banking, oil, steel, aluminum, and coal. Today, the Mellon family is still going strong, with a net worth of $14.1 billion in 2024.

922. Most Famous Pioneer Family: The Ingalls family was made famous by Laura Ingalls Wilder's "Little House" series.

29

MILITARY

923. An unusual event occurred on July 23, 1945, During World War II. Under Commander Eugene Fluckey's command, the USS Barb was on its 12th and final war patrol. The crew spotted a railway line along the coast of Karafuto Prefecture (now part of Sakhalin Island, Russia). A small team from the submarine went ashore under the cover of darkness and planted explosives on the tracks. When a train passed over the spot, the explosives detonated, destroying the locomotive and several cars. This operation made the USS Barb the only submarine in history to have destroyed a train.

924. Georg Gartner, the German soldier who pulled off a daring escape from a POW camp in New Mexico and lived under a false identity in the US for 40 years! He adopted the name Dennis Whiles and worked various jobs, from dishwasher to ski instructor, always keeping one step ahead of the FBI, who had him on their most wanted list. He even wrote a book about his experiences, titled "Hitler's Last Soldier in America."

925. Yes, indeed! Homing pigeons played a crucial role in wars; some were even awarded medals for their bravery. One famous example is Cher Ami, a homing pigeon who delivered a message that saved the lives of nearly 200 American soldiers during World War I. Cher Ami was awarded the Croix de Guerre medal by the French government for her heroic service. Another notable pigeon is Winkie, who received the Dickin Medal for rescuing a bomber crew downed in the North Sea during World War II.

926. Explosion at sea kills high-ranking officials. USS Princeton was a screw steam warship notable for an incident on February 28, 1844, This is when one of its guns, ironically named the "Peacemaker," exploded during a demonstration cruise on the Potomac River. This explosion resulted in the deaths of several high-ranking U.S. officials who were in attendance. This included the Secretary of State Abel P. Upshur and Secretary of the Navy Thomas Walker Gilmer. Due to the significant loss of life, there was a large impact on naval technology and safety protocols.

927. Another woman in history that must be mentioned. Ann A. Bernatitus was a trailblazing U.S. Navy nurse during World War II, and she was renowned for her bravery during the siege of Bataan and Corregidor. As the first American to receive the Legion of Merit, she treated soldiers from both sides amidst intense conflict and was among the last nurses evacuated before Corregidor fell. Her legacy lives on, with her Legion of Merit medal enshrined in the Smithsonian and with a monument dedicated to her in Exeter, Pennsylvania.

928. M&M's were created with a candy coating to keep the chocolate from melting in high temperatures, making them ideal for military rations during World War II. Initially, they were exclusively sold to the U.S. military.

929. The USS Stein, a Knox-class frigate of the US Navy, experienced a mysterious incident in 1978. While on patrol, the ship's sonar dome, which was coated with a special "NOFOUL" rubber, was found to be damaged by numerous cuts. These cuts covered about 8% of the dome's surface and were believed to have been caused by the claws of a giant squid. The USS Stein's encounter remains one of the more intriguing and unusual maritime mysteries.

930. The USS Cyclops: This US Navy ship disappeared in 1918 with 306 crew members and passengers while traveling from Barbados to Baltimore. No wreckage or bodies were ever found, and the cause of its disappearance remains unknown.

931. Oh, the Nth Country Experiment! That was quite the explosive project, wasn't it? In the 1960s, the US government decided to play a little game of "How easy is it to make a nuclear bomb?" They picked three bright young physicists fresh out of grad school and gave them a simple task: design a working nuclear weapon without any classified information. The result? Well, let's just say it was a blast! The physicists came up with a design that was deemed credible by the lab's weapons experts.

932. Oh boy, talk about a close call! So, back in 1980, a wrench socket fell on a Titan II missile silo in Arkansas, and the resulting explosion almost blew the whole place sky-high. I mean, we're talking about a nine-megaton warhead here, folks. That's enough to turn Arkansas into a giant crater. The story goes that a young airman was working on the missile when he

dropped the wrench socket, bouncing around like a pinball and puncturing the fuel tank. Next thing you know, the whole silo is filling up with toxic fumes, and everyone's scrambling to contain the situation before it turns into a nuclear barbecue.

933. On March 11, 1958, a U.S. Air Force B-47 Stratojet was flying over Mars Bluff, South Carolina, when a nuclear bomb accidentally fell from the aircraft. The bomb, a Mark 6 nuclear bomb, did not have its fissile core installed, so it did not cause a nuclear explosion. However, the conventional explosives in the bomb detonated upon impact, creating a large crater and causing significant damage to nearby buildings. Fortunately, there were no fatalities, but six people were injured.

934. The B-52 Stratofortress has been a loyal servant to the United States Air Force since 1955, and it doesn't seem like it's planning to retire anytime soon! This old bird is expected to keep flying high until at least the 2050s, thanks to ongoing upgrades and a whole lot of love from its maintainers.

935. On September 15, 1980, a B-52H bomber (loaded with nuclear weapons) caught fire while parked at the base. The fire started in one of the wing tanks and quickly spread, thanks to the 26-mile-per-hour wind blowing the flames toward the fuselage. Luckily, the wind stayed steady, and the fire was eventually put out after a grueling three-hour battle.

936. The first successful aerial refueling took place on June 27, 1923. Two U.S. Army Air Service DH-4B biplanes accomplished this feat by passing fuel through a hose between them. This milestone event paved the way for developing in-flight refueling techniques still used today.

937. The 25th Infantry Bicycle Corps was a unique and pioneering unit in the U.S. Army, composed of African American soldiers known as Buffalo Soldiers. In 1897, they embarked on an extraordinary journey from Missoula, Montana, to St. Louis, Missouri, covering 1,900 miles in 41 days.

938. The Ghost Army During World War II, this unit was composed of artists, designers, and sound effects experts. They aimed to deceive the enemy by creating fake military installations and movements using inflatable tanks, sound trucks, and other tricks.

939. The 1st Special Service Force, also known as the "Devil's Brigade," was a joint American-Canadian commando unit during World War II. It was trained in various unconventional warfare techniques and conducted operations in Italy and southern France.

940. The 10th Mountain Division: Originally formed during World War II, this unit specialized in mountain and winter warfare. They trained in the rugged terrain of Colorado and played a crucial role in the Italian Campaign. Recreational Skiers still use the cabins in the Colorado Rockies.

941. The Green Berets: Officially known as the United States Army Special Forces, they were established in 1952 with a primary mission of unconventional warfare. They are known for their guerrilla warfare, foreign internal defense, and counter-terrorism expertise.

942. In 1942, the Office of Strategic Services (OSS) was the precursor to the CIA, and its Operational Groups were specially trained units tasked with conducting guerrilla warfare, sabotage, and intelligence gathering behind enemy lines.

943. The Alamo Scouts were a Sixth United States Army reconnaissance unit in the Pacific Theater during World War II. They were named after the Alamo Force, a task force built around Sixth Army headquarters, under which they were organized in 1943. The unit was known for its daring missions behind enemy lines, including liberating the Cabanatuan Prisoner of War Camp in the Philippines in 1945.

944. The Women Airforce Service Pilots (WASP) were a group of civilian women pilots who flew military aircraft during World War II. They were the first women to fly for the United States military.

945. The first woman to receive the Medal of Valor was Mary Edwards Walker, who was awarded the Medal of Honor for her service during the Civil War.

946. In 1976, the United States Military Academy at West Point and the United States Naval Academy at Annapolis admitted their first female cadets and midshipmen.

947. The British-Indian Army adopted the heliograph in 1875, and it saw its first wartime use during the 1877 Jowaki Expedition. Brigadier General Nelson Miles used the American version of the heliograph, which used a fixed, square mirror, in his pursuit of Geronimo in the late 19th century.

948. Eugene Ely, a pioneering aviator, achieved the first successful landing of an airplane on an American aircraft carrier on January 18, 1911. He landed his Curtiss Pusher aircraft on a specially constructed wooden platform on the deck of the USS Pennsylvania, which was anchored in San Francisco Bay.

949. The first submarine commissioned by the U.S. Navy was the USS Holland (SS-1). It was launched on May 17, 1897, and officially commissioned on October 12, 19001.

950. The heliograph, a device that uses sunlight to transmit coded messages, has a fascinating history dating back to ancient times. Brigadier General Nelson Miles used the American version of the heliograph, which used a fixed, square mirror, in his pursuit of Geronimo in the late 19th century.

951. The U.S. Air Force's Cardboard Boat Regatta is an annual event that challenges participants to design, build, and navigate boats made entirely of corrugated cardboard. It's a test of creativity, teamwork, and ingenuity, as teams have to construct seaworthy vessels using only cardboard and duct tape.

952. During World War II, the U.S. Navy trained dolphins and sea lions for various tasks, including locating underwater mines and enemy divers. This program, known as the Navy Marine Mammal Program, began in the 1960s and continues today.

953. One particularly amusing incident involved a sea lion named Seymour. During a training exercise, Seymour was tasked with retrieving a practice mine. Instead of bringing it back, he decided to play hide-and-seek with his trainers, leading them on a merry chase around the harbor.

954. The U.S. Navy SEALs were officially established on January 1, 19621. However, their roots trace back to World War II, when the U.S. military recognized the need for covert surveillance of

landing beaches and coastal defenses. This led to the formation of the Amphibious Scout and Raider School in 19422.

955. Interestingly, an earlier instance of plane-on-plane combat occurred during the Mexican Revolution on November 30, 1913. Two American mercenaries, Dean Ivan Lamb and Phil Rader, hired by opposing sides, engaged in a unique dogfight using pistols. They fired at each other but intentionally missed, making it a rather unusual and friendly encounter.

956. The first aerial dogfight involving American pilots occurred during World War I on April 14, 1918. Two American pilots, Douglas Campbell and Alan Winslow, engaged in combat with German aircraft over the Western Front in France. They successfully shot down two German planes, marking the first aerial victory for American pilots.

30

QUIRKY TOWNS AND CITIES

957. The hottest town in the United States is Death Valley, California. Specifically, Furnace Creek in Death Valley holds the record for the highest temperature ever recorded on Earth, which was 134°F (56.7°C) on July 10, 1913.

958. A house in Beebe Plain, Vermont, straddles the US-Canada border, which has entrances on both sides, allowing residents to live in both countries simultaneously.

959. Green Bank, West Virginia! Because of its quietness, the town has been called the quietest in America. It has a powerful government telescope and is part of the United States National Radio Quiet Zone, which means cell phones and Wi-Fi are a no-go.

960. In Death Valley, the water supply primarily comes from local springs and wells. One notable source is the Ash Meadows oasis, fed by an aquifer beneath a Nevada nuclear test site.

961. The groundwater in Death Valley is derived from the base of the Funeral Formation near Furnace Creek Ranch. This water is essential for the residents and the unique plant and animal life that depend on these oases to survive.

962. The largest indoor rainforest is Lied Jungle at the Omaha Henry Doorly Zoo and Aquarium in Omaha, Nebraska.

963. The Taos Pueblo is a fascinating place where people still live in nearly 900 years old buildings.

964. The Spanish conquistadors who arrived in New Mexico in the 16th century brought a unique variety of Spanish that has since evolved into a distinct dialect known as New Mexican Spanish.

965. Michigan is home to the only floating post office in the United States! The J.W. Westcott II, a 45-foot-long tugboat, operates on the Detroit River and delivers mail to passing ships.

966. How low can you go? The lowest point in the United States is Badwater Basin in Death Valley, California, at 282 feet (86 meters) below sea level.

967. The Moving Rocks: One of Death Valley's most mysterious phenomena is the "sailing stones" of Racetrack Playa. These rocks, some weighing hundreds of pounds, appear to move across the dry lake bed on their own, leaving long trails behind them. Scientists believe the movement is caused by a combination of ice, wind, and water, but the exact mechanism remains a subject of fascination

968. The record low temperature in Fairbanks, Alaska, was -66°F (-54°C) on January 14, 1934. This remains one of the coldest temperatures ever recorded in the United States.

969. Monowi, Nebraska, is a unique and fascinating place! It's an incorporated village in Boyd County, known for having a population of just one person. The Sole Resident is Elsie Eiler, the mayor, clerk, treasurer, librarian, and bartender. Over the years, residents moved away, leaving Elsie as the last one standing. -Self-Governance Elsie handles all the administrative tasks for the village, including producing a municipal road plan to secure state funding and granting herself a liquor license.

970. Fifty-four people live in the town of Angle Inlet in Minnesota. The route requires you to drive through Canada, cross the border into Canada, and then back into the United States. This quirky geographical anomaly happened due to a mapping error during the Treaty of Paris 1783.

971. The town of Jean Lafitte, LA. is a charming little spot where the streets are paved with tales of swashbuckling as the infamous pirate Jean Lafitte and his merry band of buccaneers used the town as a hideout for their daring raids and treasure hunts.

972. Canusa Street is in Stanstead, Quebec, and Derby Line, Vermont. This street is quite unique because the US-Canada border runs right down the middle of it**. On one side of the street, you're in Canada, and on the other, you're in the United States!

973. Niihau, often called the "Forbidden Island," is fascinating, with a rich history and unique culture. In 1864, Elizabeth Sinclair, a Scottish plantation owner, purchased Niihau from King Kamehameha V for $10,000. The island has remained part of the Sinclair-Robinson family ever since.

974. The Sinclairs promised to preserve the native Hawaiian culture, and to this day, Niihau remains a place where traditional Hawaiian customs and the Hawaiian language are still practiced. Access to Niihau is highly restricted. Only guests and family members are allowed to visit, which helps preserve its unique way of life.

975. The "Zone of Death" is a fascinating and somewhat eerie area within Yellowstone National Park. This 50-square-mile region is located in the Idaho section of the park. Due to a unique legal loophole, it is theoretically possible to avoid conviction for significant crimes committed there, including murder. The Sixth Amendment of the U.S. Constitution requires that a jury for a crime be composed of residents from both the state and the federal district where the crime occurred. Since the Idaho portion of Yellowstone is uninhabited, it would be impossible to form a jury, making it theoretically impossible to hold a trial.

976. Talkeetna, Alaska, had a cat named Stubbs as its honorary mayor! Stubbs, a Manx mix, became the town's unofficial mayor in 1997 and served until his death in 2017.

977. It's not a town, but it's still of interest. Abandoned City Hall Station: This beautiful station, opened in 1904, was closed in 1945 because its curved platform was unsuitable for newer, longer trains. You can still catch a glimpse of its stunning architecture by staying on the sixth train as it loops back from the Brooklyn Bridge station.

978. Centralia, Pennsylvania, is a town living on the edge—literally. Since 1962, a coal mine fire has been burning beneath the town, turning it into a real-life version of Dante's Inferno. The fire started when the town council decided to clean up a landfill by setting it on fire. The fire quickly spread to the coal mines below; the rest is history.

979. Whittier is a small town with about 200 residents, nearly all of whom live in a 14-story building called Begich Towers. It has a post office, a grocery store, a church, a police station, a health clinic, and even a bed and breakfast!

980. Marfa, Texas! First up, the Marfa Lights. These mysterious glowing orbs have been spotted floating around the desert outside Marfa since the 1800s. Some say they're ghosts, some are aliens, and some are just car headlights. Who knows?

981. In 1985, Riverside decided to shoot for the stars and declared itself the future birthplace of Captain Kirk. They even stuck a plaque in the ground behind a barber shop to mark the spot!

982. Scottsboro, Alabama! This is the town where lost luggage goes to find a new home! The Unclaimed Baggage Center is like a treasure trove for bargain hunters and the curious. It's the only store in the country that sells the contents of unclaimed airline baggage, and it's been doing so since 1970!

983. Oh, Oak Ridge, Tennessee! The town that was so top-secret, it wasn't even on the map! During World War II, Oak Ridge was one of the three primary sites for the Manhattan Project, alongside Los Alamos, New Mexico, and Hanford, Washington. The town was built from scratch in 1942 and was home to over 75,000 workers and their families.

984. Punxsutawney, Pennsylvania, is famous for... well, a groundhog named Phil! The legend goes that if Phil sees his shadow, we're in for six more weeks of winter. Spring is just around the corner if he doesn't see his shadow.

985. Roswell, New Mexico, is a small town that gained international fame due to an event in 1947. This incident, often referred to as the "Roswell UFO Incident," has become one of the most famous and controversial UFO events in history.

986. Slab City, often called "The Last Free Place in America," is a unique, off-the-grid community in the Sonoran Desert of Southern California. Slab City gets its name from the concrete slabs from the World War II Marine Corps Camp Dunlap. The residents, known as "Slabbers," live in various makeshift homes, including trailers, tents, and repurposed vehicles.

987. Gibsonton, FL, where the circus never really left town. Known as the "circus capital of the world," this town has a rich history of hosting circus performers during the off-season. It's like the Cirque du Soleil of retirement communities, minus the acrobats and a lot more Florida sun.

988. Truth or Consequences, NM: The town that took a radio show's dare and ran with it. In 1950, Hot Springs, NM, changed its name to Truth or Consequences to honor the popular radio show of the same name.

989. Nameless, TN: The town that took the phrase "What's in a name?" a bit too literally. It's like they got to the part in the town-naming process where they had to fill out the paperwork and said, "You know what? Let's leave it blank. It'll be fine." And thus, the town of Nameless was born.

990. Holy Land USA: Situated in Waterbury, Connecticut, selected passages from the Bible inspired this 18-acre park. It includes replicas of catacombs, Israelite villages, and other biblical scenes constructed from various materials. However, it was not profitable, so it closed.

991. Sturgis, South Dakota's population is approximately 7,020, according to the 2020 census. Sturgis is well-known for hosting the annual Motorcycle Rally, with as many as 700,000 motorcycle enthusiasts worldwide.

992. The Lied Jungle at the Henry Doorly Zoo in Omaha, Nebraska, is an indoor rainforest. It's the largest indoor rainforest in North America, covering 1.5 acres and featuring a variety of tropical plants and animals[2].

993. There is a Spoon Museum in New Jersey! This museum houses the world's most extensive collection of spoons, with over 5,400 from around the globe.

994. Tucson, Arizona, is renowned for its telescopes and astronomical research facilities. The city is home to several significant observatories and has earned a reputation as one of the best places for stargazing in the world.

995. Tucson, Arizona, is indeed renowned for its telescopes and astronomical research facilities. The city is home to several major observatories and has earned a reputation as one of the best places for stargazing in the world.

996. The small town of Lajitas, Texas, elected a goat named Clay Henry as its mayor in 198612. Clay Henry was not just any goat; he was known for his love of beer and became quite a local celebrity.

997. The Lost Colony of Roanoke: The mysterious disappearance of an entire colony in the late 16th century. Founded in 1585 on Roanoke Island in what is now North Carolina, the colony lost over 100 settlers, and no one knows where.

998. Havre de Grace, Maryland, proudly holds the "Decoy Capital of the World" title. The Havre de Grace Decoy Museum showcases an impressive collection of working and decorative Chesapeake Bay decoys.

31

CONSPIRACIES AND MYSTERIES

999. She was arguably the most famous aviator, Amelia Earhart. In 1937, she disappeared over the Pacific Ocean during an attempt to circumnavigate the globe, never to be found.

1000. The Philadelphia Experiment is a conspiracy theory about a supposed military experiment at the Philadelphia Naval Shipyard in 1943, where the USS Eldridge was rendered invisible and teleported. While widely debunked, it remains a famous odd story.

1001. The Georgia Guidestones: Erected in 1980, this mysterious granite monument in Georgia has inscriptions in multiple languages and unknown origins. And if that is not odd enough, it was destroyed by an explosion in July 2022.

1002. The Gardner Museum Heist: In 1990, two thieves stole 13 art pieces worth $500 million from the Isabella Stewart Gardner Museum in Boston. The artwork has never been recovered.

1003. The Montauk Project: Alleged secret experiments at Montauk Air Force Station in New York, including time travel and mind control, have inspired numerous conspiracy theories. This sounds awesome, but I can't find anyone to verify it.

1004. The Death of Tupac Shakur: The 1996 murder of the rapper in Las Vegas remains unsolved, with numerous conspiracy theories about who was responsible.

1005. The 1981 drowning of actress Natalie Wood remains a mystery, with theories suggesting foul play. She was on a yacht and had argued with her husband the night before, but no charges were ever filed. No one knows what happened that night.

1006. The Black Dahlia Murder, a brutal 1947 murder of Elizabeth Short in Los Angeles, remains an infamous unsolved case in American history. Over the years, many suspects have been named, but none have been conclusively linked to the crime.

1007. The Benders of Kansas! The original "family that slays together stays together." This family of serial killers operated an inn and general store in Kansas in the 1870s. They disappeared before they could be brought to justice.

1008. The Death of Kurt Cobain: The 1994 death of the Nirvana frontman was ruled a suicide, but some think that it is suspicious and believe he was murdered.

1009. The 1963 assassination of President John F. Kennedy has led to countless conspiracy theories questioning whether Lee Harvey Oswald acted alone. We are still waiting for all the files to be declassified.

1010. The Bermuda Triangle: This region in the North Atlantic Ocean is infamous for the mysterious disappearances of ships and aircraft.

1011. In 1918, the USS Cyclops vanished without a trace in the Bermuda Triangle; the ship, carrying 306 crew members and a cargo of manganese ore, was last seen on March 4, 1918, leaving Barbados. It was supposed to arrive in Baltimore on March 13 but never did. No distress signals were sent, no wreckage was found, and no survivors were located.

1012. The 1971 death of Jim Morrison, The Doors frontman, was ruled heart failure, but some want to believe he faked his death. Like Elvis?

1013. Jimmy Hoffa (who was not well-liked), the labor union leader, vanished in 1975, and despite numerous theories, his fate remains unknown.

1014. The Death of Bob Marley: The 1981 death of the reggae legend was officially due to cancer, but some believe he was poisoned.

1015. The Disappearance of the Sodder Children: In 1945, five children disappeared after a fire destroyed their home in West Virginia. Despite various theories, their fate remains unknown.

1016. While John Wilkes Booth is known as the assassin of President Abraham Lincoln in 1865, theories suggest a broader conspiracy involving other high-ranking officials—one more similarity to JFK.

1017. In 1897, Elva Zona Heaster Shue was found dead at the bottom of the stairs in her home in Greenbrier County, West Virginia. Her husband, Erasmus "Trout" Shue, was charged with her murder after her mother, Mary Jane Heaster, claimed that her daughter's ghost visited her in dreams and told her that Trout had killed her.

1018. In 1872, the American merchant ship Mary Celeste was adrift and deserted in the Atlantic Ocean. The fate of its crew remains a mystery.

1019. After the Civil War, rumors spread that the Confederacy's gold reserves were hidden or lost. Various treasure hunters have searched for it, but its location remains unknown.

1020. The Oak Island Money Pit, located on the east side of Oak Island, Nova Scotia, is a legendary treasure-hunting site that has been the focus of numerous expeditions since its discovery in 1795. The Money Pit is a deep shaft, originally over 100 feet deep, featuring wooden platforms at 10-foot intervals. Over the years, treasure hunters have tried to uncover what they believe to be a vast treasure buried beneath the island.

32

SOURCE

Mythologies

Mythologies of the Indigenous peoples of the Americas.
https://en.wikipedia.org/wiki/Mythologies_of_the_indigenous_peoples_of_the_Americas.

Native American Mythology 101: The Ultimate Guide - MythBank. https://mythbank.com/native-american-mythology/

Native American Mythology: Legends of the First People. https://mythologis.com/blogs/native-american-mythology/native-american-mythology-introduction.

Indigenous Myths: Native American Mythology. https://mythosaurus.com/native-american-mythology.

Native American Myths & Folklore - mythosaurus.com. https://mythosaurus.com/native-american-myths.

Our Long History of Bigfoot Stories & Accounts – Bigfoot Blog. https://bigfootblog.com/history-of-bigfoot-accounts/.

Founding fathers

20 Intriguing Facts about the Revolutionary War - Discover Walks.
https://www.discoverwalks.com/blog/united-states/20-intriguing-facts-about-the-revolutionary-war/.

30 Fun And Interesting Facts About The American Revolution. http://tonsoffacts.com/30-fun-interesting-facts-american-revolution/.

45 Things You Definitely Didn't Know About The Founding Fathers.
https://www.buzzfeed.com/awesomer/founding-father-facts.

Did the Founding Fathers Have Pets? - Barks & Blooms. https://barksandblooms.com/did-the-founding-fathers-have-pets/.

19 Little-Known Facts About America's Founding Fathers.
https://explorethearchive.com/founding-fathers-facts.

Us Presidents

30 Craziest Things U.S. Presidents Have Done — Best Life. https://bestlifeonline.com/craziest-things-presidents/.

45 Odd Facts About U.S. Presidents - Mental Floss. https://www.mentalfloss.com/article/49694/45-odd-facts-about-us-presidents.

Strange, Fun and Weird Facts about US Presidents. https://constitutionus.com/presidents/unusual-facts-about-us-presidents/.

15 Quirks of U.S. Presidents You Didn't Learn in School. https://www.mentalfloss.com/article/61593/15-quirks-us-presidents-you-didnt-learn-school.

The Secret Lives Of U.S. Presidents - All That's Interesting. https://allthatsinteresting.com/history-uncovered/facts-about-american-presidents.

Landmarks:

(10 fascinating facts about the Washington Monument. https://constitutioncenter.org/blog/10-fascinating-facts-about-the-washington-monument.

30+ FASCINATING Facts About World Landmarks - Travel Trivia Challenge. https://traveltriviachallenge.com/world-landmarks-facts/.

494 Unusual Monuments in the United States - Atlas Obscura. https://www.atlasobscura.com/things-to-do/united-states/monuments.

Top 51 Unusual Places to Visit in the USA That Will Amaze You. https://www.attractionsofamerica.com/thingstodo/top-10-unusual-places-to-visit-in-the-usa.php.

11 Secret Spaces Hiding in Famous Places - Atlas Obscura. https://www.atlasobscura.com/articles/11-secret-spaces-hiding-in-famous-places.

Hidden Secrets of the World's Most Famous Landmarks - Life123.com. https://www.life123.com/lifestyle/Hidden-Secrets-Worlds-Most-Famous-Landmarks.

Secrets behind famous monuments - Stars Insider. https://www.starsinsider.com/travel/339646/secrets-behind-famous-monuments.

laws

10 of the strangest laws in New York State - silive.com. https://www.silive.com/news/2017/05/strangest_laws_in_new_york_sta.html.

These 10 Weird Laws In New York Are Unbelievable - Only In Your State. https://www.onlyinyourstate.com/new-york/crazy-laws-ny/.

11 of the Strangest Laws in NYC and New York State.
https://untappedcities.com/2017/09/28/cities-101-10-of-the-strangest-new-york-city-state-laws/2/.

These 10 Weird Laws In New York Are Unbelievable - Only In Your State.
https://www.onlyinyourstate.com/new-york/crazy-laws-ny/.

15 Ridiculous Laws In New York State You're Probably Breaking - 106.5 .... https://wyrk.com/15-ridiculous-laws-in-new-york-state-youre-probably-breaking/

26 Weird Laws You Won't Believe Existed In The U.S. - BuzzFeed.
https://www.buzzfeed.com/rhiannacampbell/weird-old-american-laws-you-wont-believe.

Top craziest laws still on the books | LegalZoom. https://www.legalzoom.com/articles/top-craziest-laws-still-on-the-books.

Civil War

[History.com - Medical Innovations of the Civil War](https://www.history.com/topics/american-civil-war/civil-war-medicine)

[Smithsonian Magazine - Civil War Medical Practices](https://www.smithsonianmag.com/history/civil-war-medicine-180964568/)

[History.com - Transcontinental Railroad](https://www.history.com/topics/inventions/transcontinental-railroad)

60 US Civil War Trivia Questions And Answers - Land of Trivia. https://landoftrivia.com/civil-war-trivia/.

U.S. Civil War Trivia. http://usefultrivia.com/war_trivia/civil_war_trivia_index.html.

Ten things you didn't know about the Civil War - Pieces of History.
https://prologue.blogs.archives.gov/2010/11/01/ten-things-you-didnt-know-about-the-civil-war/.

7 Civil War Stories You Didn't Learn in High School.
https://www.mentalfloss.com/article/21936/7-civil-war-stories-you-didnt-learn-high-school.

Last surviving United States war veterans - Wikipedia.
https://en.wikipedia.org/wiki/Last_surviving_United_States_war_veterans.

Civil War Casualties | American Battlefield Trust.
https://www.battlefields.org/learn/articles/civil-war-casualties.

The California Gold Rush | American Experience | PBS.
https://www.pbs.org/wgbh/americanexperience/features/goldrush-california/.

Going for Gold: How the Confederacy Hatched an Audacious Plan to ....
https://www.historynet.com/confederate-plan-to-finance-war/.

List of naval battles of the American Civil War - Wikipedia.
https://en.wikipedia.org/wiki/List_of_Naval_battles_of_the_American_Civil_War.

Native American

20 Famous Native Americans in American History. https://thehistoryjunkie.com/20-famous-native-americans-in-american-history/.

13 Most Famous Native Americans - Have Fun With History. https://www.havefunwithhistory.com/famous-native-americans/.

Greatest Native Americans of All Time - World History Edu. https://worldhistoryedu.com/greatest-native-americans-of-all-time-and-achievements/.

20 Famous Native Americans We Should All Know About. https://www.discoverwalks.com/blog/united-states/20-famous-native-americans-we-should-all-know-about/.

Native American History Timeline. https://www.history.com/topics/native-american-history/native-american-timeline.

Native American Historymakers to Know | TIME. https://time.com/6317481/native-american-history-makers/.

Native American History: Tribes, Timeline & Reservations | HISTORY. https://www.history.com/topics/native-american-history.

A Guide to Native American Pottery - The Spruce Crafts. https://www.thesprucecrafts.com/native-american-pottery-4157700.

History of Native Americans in the United States - Wikipedia. https://en.wikipedia.org/wiki/History_of_Native_Americans_in_the_United_States.

94 Interesting Native American Facts | Fact Retriever. https://www.factretriever.com/native-american-facts.

Ten North American Native Facts You Need To Know. https://www.worldhistory.org/article/2347/ten-north-american-native-facts-you-need-to-know/.

Pioneers and Explorers

First Emigrants on the Oregon Trail - OCTA. https://octa-trails.org/articles/first-emigrants-on-the-oregon-trail/.

A thousand pioneers head West as part of the Great Emigration. https://www.history.com/this-day-in-history/a-thousand-pioneers-head-west-on-the-oregon-trail.

Oregon Trail, History, Facts, Significance, Summary, APUSH. https://www.americanhistorycentral.com/entries/oregon-trail/.

1843 - Year of the Great Migration | Savages & Scoundrels. http://www.savagesandscoundrels.org/events-landmarks/1843-year-of-the-great-migration/.

First Emigrants on the Oregon Trail - OCTA. https://octa-trails.org/articles/first-emigrants-on-the-oregon-trail/.

A thousand pioneers head West as part of the Great Emigration. https://www.history.com/this-day-in-history/a-thousand-pioneers-head-west-on-the-oregon-trail.

Eleven Famous Women Immigrants in the United States. https://www.ilctr.org/famous-women-immigrants/.

10 Facts About Pioneers of Western America - Have Fun With History. https://www.havefunwithhistory.com/facts-about-pioneers/.

Outfitting for the Journey – End of the Oregon Trail. https://historicoregoncity.org/2019/04/03/outfitting-for-the-journey/.

Mules, Horses or Oxen - Learn what Pioneers used to pull covered wagons. https://oregontrailcenter.org/mules-oxen.

Wild west

Cattle drives in the United States - Wikipedia. https://en.wikipedia.org/wiki/Cattle_drives_in_the_United_States.

[hobbylark.com - American Wild West Quiz (With Answers)](https://hobbylark.com/party-games/Free-Wild-West-Quiz)

Tombstone | Smithsonian. https://www.smithsonianmag.com/history/tombstone-117086409/.

The Ghosts of Tombstone - Tucson Community Guide. https://tucsonshiddengem.com/tombstone-ghosts/.

History of Wells Fargo – Wells Fargo. https://www.wellsfargo.com/about/corporate/history/.

Riding shotgun - Wikipedia. https://en.wikipedia.org/wiki/Riding_shotgun.

Riding Shotgun - Meaning & Origin Of The Phrase - Phrasefinder. https://www.phrases.org.uk/meanings/riding-shotgun.html.

10 Wild West Facts of Everyday Life on the Frontier - OldWest. https://bing.com/search?q=fun+fact+about+Old+West+water+sources.

10 Wild West Facts of Everyday Life on the Frontier - OldWest. https://www.oldwest.org/wild-west-facts/.

25 Amazing Facts You Didn't Know About the Wild West. https://www.farandwide.com/s/wild-west-facts-099d473363fa4494.

History of Water in the West | ORCA. https://archaeologycolorado.org/resources.

Forts of the Old West - North Pole West. https://www.northpolewest.com/Forts-of-the-Old-West_ep_88.html.

Lawmen/women

10 Most Famous Lawmen of the Old West - Have Fun With History.
https://www.havefunwithhistory.com/lawmen-of-the-old-west/.

25 of the Most Influential Women in American History - The Daily Signal.
https://www.dailysignal.com/2018/03/28/25-of-the-most-influential-women-in-american-history/.

10 Female Lawyers Who Shaped American History - PracticePanther.
https://www.practicepanther.com/blog/female-lawyers-american-history/.

List of Old West lawmen - Wikipedia. https://en.wikipedia.org/wiki/List_of_Old_West_lawmen.

The Major Events and Women that Shaped U.S. Legal History.
https://messerlikramer.com/women-that-shaped-us-legal-history/.

Women's History Month: First women lawyers, judges around the world.
https://law.duke.edu/news/womens-history-month-first-women-lawyers-judges-around-world.

A Timeline of Women's Legal History in the United States and at ....
https://wlh.law.stanford.edu/wp-content/uploads/2011/01/cunnea-timeline2.pdf

Trailblazing U.S. Law Women | Mental Floss. https://www.mentalfloss.com/article/83601/10-trailblazing-us-law-women.

50 years ago, women made history at FBI: "I certainly wasn't going to ....
https://www.cbsnews.com/news/fbi-first-female-special-agents-susan-roley-malone/.

Odd laws

50 Weird Laws in the U.S | Far & Wide. https://www.farandwide.com/s/weird-laws-united-states-5ec88a12367547fd.

The 38 Weirdest Laws From Across the U.S. — Best Life. https://bestlifeonline.com/weirdest-laws/.

Weirdest laws passed in every state - USA TODAY. https://www.usatoday.com/list/news/nation-now/weirdest-laws-every-state/53ad0541-3518-4432-adc4-0fec193d389e/

The Most Ridiculous Laws in 30 American States. https://inspiredbyinsiders.com/the-most-ridiculous-laws/.

4 Creepy Unsolved Mysteries In Alaska That Will Leave You Baffled.
https://www.onlyinyourstate.com/alaska/disturbing-unsolved-mysteries-in-ak/.

The Craziest Laws That Still Exist In The United States. https://www.huffpost.com/entry/weird-laws-in-america_n_56a264abe4b0d8cc1099e1cd.

Weird, obsolete laws exist in the hundreds. Why don't more ... - Vox. https://www.vox.com/the-highlight/2019/11/18/20963411/weird-old-laws-historical-obsolete-laws.

The Weirdest Animal Laws In The US - I Can Has Cheezburger?.
https://cheezburger.com/4120837/the-weirdest-animal-laws-in-the-us.

Legal Stuff

Detailed Timeline - National Women's History Alliance.
https://nationalwomenshistoryalliance.org/resources/womens-rights-movement/detailed-timeline/.

Evolution of the Legal Profession in the USA: A Timeline.
https://americanprofessionguide.com/usa-legal-profession-evolution/.

Protection - United States Secret Service. https://www.secretservice.gov/protection.

History - United States Secret Service. https://www.secretservice.gov/about/history.

Timeline of Our History - United States Secret Service.
https://www.secretservice.gov/about/history/timeline.

150+ Years of History - United States Secret Service.
https://www.secretservice.gov/about/history/150-years.

Crimes/ outlaws

9 Things You May Not Know About Bugsy Siegel | HISTORY. https://www.history.com/news/9-things-you-may-not-know-about-bugsy-siegel.

How Hardened Gangsters Got the Cute Name 'Bugsy' - Atlas Obscura.
https://www.atlasobscura.com/articles/how-hardened-gangsters-got-the-cute-name-bugsy.

Criminal law of the United States - Wikipedia.
https://en.wikipedia.org/wiki/Criminal_law_of_the_United_States.

criminal law | Wex | US Law | LII / Legal Information Institute.
https://www.law.cornell.edu/wex/criminal_law.

The Justice System | Bureau of Justice Statistics. https://bjs.ojp.gov/justice-system.

The 10 Biggest Bank Robberies of All Time | Moneywise.
https://moneywise.com/life/entertainment/the-biggest-bank-robberies-of-all-time.

D. B. Cooper - Wikipedia. https://en.wikipedia.org/wiki/D._B._Cooper.

When Do Most Burglaries Occur? Everything You Should Know - ADT Security.
https://www.adt.com/resources/when-do-most-burglaries-occur.

The Most Common Time of Day for Burglaries | Reader's Digest.
https://www.rd.com/article/most-common-time-for-burglaries/.

Crime in the U.S.: Key questions answered | Pew Research Center.
https://www.pewresearch.org/short-reads/2024/04/24/what-the-data-says-about-crime-in-the-us/.

Interesting Building

18 Most Famous Buildings and Monuments in the USA - Time Out.
https://www.timeout.com/usa/things-to-do/famous-buildings-in-america.

30 of the Best Historic Sites in the United States. https://www.historyhit.com/guides/historic-sites-in-the-united-states/.

Top 10 Most Interesting Buildings in America.
https://www.awardwinningdestinations.com/info/explore/top-10-most-interesting-buildings-in-america/.

Historic Virginia Homes - Virginia Is For Lovers. https://www.virginia.org/things-to-do/history-and-heritage/historic-homes/.

10 stately homes still lived in by the descendants you can visit.
https://www.telegraph.co.uk/family/life/10-stately-homes-still-lived-descendants-

10 Oldest Homes In America (With Pictures) in 2024. https://a-z-animals.com/blog/oldest-homes-in-america-with-pictures/.

America's smallest house is just one of many spite houses, built to ....
https://www.homesandgardens.com/news/americas-smallest-house-spite-house.

Small but perfectly formed: Top 10 tiny houses of 2023 - New Atlas. https://newatlas.com/tiny-houses/top-10-tiny-houses-2023/.

1123 Places to Experience Unusual Architecture in the United States.
https://www.atlasobscura.com/things-to-do/united-states/architecture.

10 modernist architectural marvels on America's East Coast - Dezeen.
https://www.dezeen.com/2018/10/08/10-modernist-architectural-marvels-east-

Geography

List of mountain peaks of the United States - Wikipedia.
https://en.wikipedia.org/wiki/List_of_mountain_peaks_of_the_United_States.

The 12 Tallest Mountains in the United States - Treehugger. https://www.treehugger.com/tallest-mountains-in-the-united-states-5184962.

U.S. 14ers By State - Infographic – 14Air. https://www.14air.net/blogs/colorado-14er-stories/u-s-14ers-by-state-infographic.

Prime meridian - Wikipedia. https://en.wikipedia.org/wiki/Prime_meridian.

Greenwich meridian | Definition, History, Location, Map, & Facts. https://www.britannica.com/place/Greenwich-meridian.

The Top 9 Highest Paved Roads in the United States. https://snowbrains.com/the-top-9-highest-paved-roads-in-the-united-states/.

Trail Ridge Road Scenic Drive in Rocky Mountain National Park. https://www.rockymountainnationalpark.com/gallery/drive-trr/.

United States: lowest point in each state | Statista. https://www.statista.com/statistics/1325443/lowest-points-united-states-state/.

List of lakes of Colorado - Wikipedia. https://en.wikipedia.org/wiki/List_of_lakes_of_Colorado.

Colorado Fourteeners (CO) | Fastest Known Time. https://fastestknowntime.com/route/colorado-fourteeners-co.

Mammoth Cave National Park - Wikipedia. https://en.wikipedia.org/wiki/Mammoth_Cave_National_Park.

Animals

Weird Animal-Related Laws In The U.S. - Critter Culture. https://critterculture.com/resources/weird-animal-related-laws-in-the-u-s/.

History of bear wrestling and when it was outlawed - The Courier. https://www.houmatoday.com/story/special/2019/08/11/history-of-bear-wrestling-and-when-it-was-outlawed/4490860007/.

7 of the World's Most Dangerous Lizards and Turtles. https://www.britannica.com/list/7-of-the-worlds-most-dangerous-lizards-and-turtles.

35 Fun Facts About Roadrunners That Will Blow Your Mind!. https://learnbirdwatching.com/fun-facts-about-roadrunners/.

What's the Deadliest Snake in the World? | Field & Stream. https://www.fieldandstream.com/story/survival/the-worlds-deadliest-snakes/.

Praying Mantis Fun Facts – 10 Mantis Facts You Probably Didn't Know. https://praying-mantis.org/praying-mantis-fun-facts/.

Complex Communication**: Prairie dogs have one of the most sophisticated animal languages known. Their calls can convey detailed information about predators, including size, shape, color, and even the speed of approach[12].

18 Amazing Animals Only Found in the United States. https://exploring-usa.com/animals-only-found-united-states/.

The eyes of a donkey are positioned so that it can see all four feet at ....
https://thefactbase.com/the-eyes-of-a-donkey-are-positioned-so-that-it-can-see-all-four-feet-at-all-times/.

Glow-in-the-Dark Animals: Various animals, including mice, rabbits, and fish, have been genetically modified to glow under UV light by inserting genes from jellyfish or other bioluminescent organisms[2].

Faith, the two-legged dog who inspired millions, dies surrounded by ....
https://animalwellnessmagazine.com/faith-two-legged-dog-dies/.

5 Beautiful Species You Can Only Find in Alaska and How to ... - Matador. Https:/

/matadornetwork.com/life/5-beautiful-species-can-find-alaska-help-thrive/.

Transportation

At 316.11 MPH, the 2020 SSC Tuatara Hypercar Is Now the World's Fastest ....
https://www.motortrend.com/news/ssc-tuatara-worlds-fastest-production-car-record/.

In 1894, the first bike lane in America was built on Brooklyn's Ocean ....
https://www.6sqft.com/in-1894-the-first-bike-lane-in-america-was-built-on-brooklyns-ocean-parkway/.

[History.com - Early Automobile Regulations](https://www.history.com/topics/inventions/automobiles)

[Smithsonian Magazine - The First Automobile Law in America](https://www.smithsonianmag.com/innovation/the-first-automobile-law-in-america-140493565/)

An Early Airliner That Cost $65,000: The Boeing 247 - Simple Flying.
https://simpleflying.com/boeing-247-story-and-cost/.

Boeing 247 - Wikipedia. https://en.wikipedia.org/wiki/Boeing_247.

Boeing 247 - Specifications - Technical Data / Description.
https://www.flugzeuginfo.net/acdata_php/acdata_boeing_247_en.php.

Traveling the Emigrant Trails - U.S. National Park Service.
https://www.nps.gov/articles/000/traveling-emigrant-trails.htm.

The world's 10 longest railway networks - Railway Technology. https://www.railway-technology.com/features/featurethe-worlds-longest-railway-networks-4180878/.

Pop culture / Cultural Icons

Top 100 Celebrity Facts - The Fact Site. https://www.thefactsite.com/100-funny-celebrity-facts/.

59 Weird Facts About Celebrities - BuzzFeed. https://www.buzzfeed.com/mjs538/strange-celebrity-facts.

35 Obscure, Interesting And Bizarre Facts About Celebrities That Aren't ....
https://www.boredpanda.com/random-interesting-celebrities-facts/.

Weird Celebrity Facts That Are 100% True - Ranker. https://www.ranker.com/list/weird-but-true-celebrity-facts/katia-kleyman.

List of Academy Award records - Wikipedia.
https://en.wikipedia.org/wiki/List_of_Academy_Award_records.

Who has the most Oscars? Top record-holders in Academy Awards history..
https://www.usatoday.com/story/entertainment/movies/oscars/2023/03/03/who-has-most-oscars/11365523002/.

27 Movies With The Most Oscars Won In History - Variety. https://variety.com/lists/movies-most-oscars-won/.

Actor Johnny Depp turns to tarot to inspire art collection.
https://www.hindustantimes.com/entertainment/tv/actor-johnny-depp-turns-to-tarot-to-inspire-art-collection-101721257369509.html.

Rudolph the Red-Nosed Reindeer - Wikipedia. https://en.wikipedia.org/wiki/Rudolph_the_Red-Nosed_Reindeer.

Food

100 Mind-Blowing Facts About Food — Eat This Not That. https://www.eatthis.com/mind-blowing-food-facts/.

American Food Favorites: 30 Must Eat Dishes - 2foodtrippers.
https://www.2foodtrippers.com/american-food-favorites/.

31 Surprising Food Facts You'll Want to Know - Reader's Digest.
https://www.rd.com/article/food-facts-trivia/.

History of Fast-food restaurant in Timeline - Popular Timelines.
https://populartimelines.com/timeline/Fast-food-restaurant.

The Real Colonel Sanders Hated Everything that KFC Became - Food & Wine.
https://www.foodandwine.com/comfort-food/real-colonel-sanders-hated-everything-kfc-became.

Potato chip - Wikipedia. https://en.wikipedia.org/wiki/Potato_chip.

The Curious History of the Potato Chip | Smithsonian. https://www.smithsonianmag.com/arts-culture/curious-history-potato-chip-180979232/.

The Origin Stories of 25 of Your Favorite Fast Food Chains.
https://www.mentalfloss.com/article/514254/origin-stories-25-your-favorite-fast-food-chains.

Sanders, Harland David ("Colonel") | Encyclopedia.com.
https://www.encyclopedia.com/humanities/encyclopedias-almanacs-transcripts-and-maps/sanders-harland-david-colonel.

Reuben Sandwich: the best recipe for a popular American sandwich.
https://www.cookist.com/reuben-sandwich/p2/?ref=shortener.

Classic American Grilled Cheese. https://www.foodnetwork.com/recipes/jeff-mauro/classic-american-grilled-cheese-recipe-2124785.

National Peanut Butter and Jelly Day – April 2, 2025. https://nationaltoday.com/national-peanut-butter-and-jelly-day/.

Sports

CU professor explains the science behind Coors Field homers - Denver7.
https://www.denver7.com/news/local-news/the-science-behind-coors-field-homers.

The Physics of Coors Field's Higher Fence | The Hardball Times. https://tht.fangraphs.com/the-physics-of-coors-fields-higher-fence/.

The Greatest Left Handed Baseball Players of All Time - Ranker.
https://www.ranker.com/list/greatest-left-handed-baseball-players-of-all-time/ranker-

The Origins of 7 Popular Sports | HISTORY. https://www.history.com/news/origins-invention-popular-sports.

A Brief History and the Evolution of Sports in the United States. https://brewminate.com/a-brief-history-and-the-evolution-of-sports-in-the-united-states/.

How the U.S. became a sporting culture | Sporting News.
https://www.sportingnews.com/us/more/news/how-the-us-became-a-sporting-culture/1vgv4kxl2459w1l6dvrbhev2er.

History of sports in the United States - Wikipedia.
https://en.wikipedia.org/wiki/History_of_sports_in_the_United_States.

Odd

If Identical Twins Married Identical Twins, How Genetically Similar ....
https://www.livescience.com/63382-identical-twins-marriage-children.html.

https://www.mentalfloss.com/posts/history-greatest-mysteries.

7 Best Mystery Books to Read Right Now (According to Mystery Experts) - PBS.
https://www.pbs.org/wgbh/masterpiece/specialfeatures/7-best-mystery-books-according-to-mystery-experts/.

30 Unsolved Mysteries that Fascinate Americans — Best Life.
https://bestlifeonline.com/unsolved-mysteries/.

15 of History's Greatest Mysteries - Mental Floss.

Medical/ Health

8 old-fashioned medical remedies that are still being used.
https://wexnermedical.osu.edu/blog/old-fashioned-medicine-still-in-use.

These Surprising Items Were Used As Cures In The Wild West.
https://www.grunge.com/649791/these-surprising-items-were-used-as-cures-in-the-wild-west/.

How A Gunshot Wound Cured One Man's OCD - KnowledgeNuts.
https://knowledgenuts.com/2014/10/17/how-a-gunshot-wound-cured-one-mans-ocd/.

8 Unsolved Medical Mysteries That Still Stump Doctors. https://www.rd.com/list/unsolved-medical-mysteries-still-stump-doctors/.

Medical Contributions by Native Americans That Are Used Every Day ....
https://www.adventhealth.com/blog/8-medical-contributions-native-americans-are-used-every-day.

13 Black American Pioneers Who Changed Healthcare - Everyday Health.
https://www.everydayhealth.com/healthy-living/african-american-pioneers-who-changed-healthcare/.

Celebrating 10 African-American medical pioneers | AAMC.
https://www.aamc.org/news/celebrating-10-african-american-medical-pioneers.

Innovation

History of the bicycle - Wikipedia. https://en.wikipedia.org/wiki/History_of_the_bicycle.

origin of the Bicycle: A Brief History of Its Invention. https://bicyclepotential.org/blog/when-was-the-bicycle-invented-a-look-back-at-the-origins-of-cycling.

Why Does America Prize Creativity and Invention? | Smithsonian.
https://www.smithsonianmag.com/innovation/why-does-america-prize-creativity-and-invention-180957256/.

Advancing American Innovation in the National Interest - Aspen Institute.
https://www.aspeninstitute.org/wp-content/uploads/2022/03/American-Innovation_Final.pdf.

American Innovation $1 Coin Program | U.S. Mint. https://www.usmint.gov/learn/coin-and-medal-programs/american-innovation-dollar-coins.

https://bing.com/search?q=American+innovation.

Home : American Innovations. https://www.aiworldwide.com/.

Science

8 Native American Scientists And Their Important Contributions. https://www.discovermagazine.com/the-sciences/8-native-american-scientists-you-should-know.

8 Native American Scientists You Should Know | HowStuffWorks. https://science.howstuffworks.com/dictionary/famous-scientists/physicists/native-american-scientists.htm.

Native American Scientists and Engineers - Science Buddies. https://www.sciencebuddies.org/blog/native-american-scientists-engineers.

Native Americans and Science | Native American Heritage Month. https://scienceatl.org/native-knowledge-and-next-steps-in-science/.

Spider-Man Can: USU Scientist says Synthetic ... - Utah State University. https://www.usu.edu/today/story/spider-man-can-usu-scientist-says-synthetic-spider-silk-research-advancing.

7 genetically modified animals that glow in the dark. https://theweek.com/articles/464980/7-genetically-modified-animals-that-glow-dark.

7 Genetically Modified Animals That Glow in the Dark. https://www.mentalfloss.com/article/50448/7-genetically-modified-animals-glow-dark.

The World's First Solar-Powered Satellite is Still Up There After More .... https://www.smithsonianmag.com/smart-news/worlds-first-solar-powered-satellite-still-there-after-59-years-180962510/.

The Ten Most Significant Science Stories of 2023 | Smithsonian. https://www.smithsonianmag.com/science-nature/the-ten-most-significant-science-stories-of-2023-180983484/.

The Top Ten Scientific Discoveries of the Decade | Smithsonian. https://www.smithsonianmag.com/science-nature/top-ten-scientific-discoveries-decade-180973873/.

75 Breakthroughs by America's National Laboratories. https://www.energy.gov/articles/75-breakthroughs-americas-national-laboratories-0.

10 Popular Scientific Discoveries From 2021 | Smithsonian Voices .... https://www.smithsonianmag.com/blogs/national-museum-of-natural-history/2021/12/28/10-popular-scientific-discoveries-from-2021/.

These Are The Biggest Scientific Discoveries in Every US State. https://www.sciencealert.com/these-are-the-biggest-scientific-discoveries-in-every-us-state.

First Clinical X-ray in America Performed | Dartmouth. https://home.dartmouth.edu/about/first-clinical-x-ray-america-performed.

Space Exploration

NASA: 60 Years and Counting - Human Spaceflight.
https://www.nasa.gov/specials/60counting/spaceflight.html.

NASA's 10 Greatest Achievements | HowStuffWorks. https://science.howstuffworks.com/ten-nasa-achievements.htm.

60 Moments in NASA History. https://www.nasa.gov/specials/timeline/.

UNITED STATES SPACE PRIORITIES FRAMEWORK | The White House.
https://www.whitehouse.gov/briefing-room/statements-releases/2021/12/01/united-states-space-priorities-framework/.

List of space programs of the United States - Wikipedia.
https://en.wikipedia.org/wiki/List_of_space_programs_of_the_United_States.

100 Interesting Space Facts That'll Blow Your Mind. https://bing.com/search?q=unique+space-related+facts.

Space Facts - 50 Things You Probably Didn't Know - The Planets. https://theplanets.org/space-facts/.

100 Interesting Space Facts That'll Blow Your Mind. https://www.thefactsite.com/100-space-facts/.

Space Facts - Interesting Facts about Space!. https://space-facts.com/.

10 Facts about Space! - National Geographic Kids.
https://www.natgeokids.com/uk/discover/science/space/ten-facts-about-space/.

Weather

Lowest Temperatures in Fairbanks History - Extreme Weather Watch.
https://www.extremeweatherwatch.com/cities/fairbanks/lowest-temperatures.

Fairbanks Weather Records - Extreme Weather Watch.
https://www.extremeweatherwatch.com/cities/fairbanks.

Hottest Cities in United States - Current Results. https://www.currentresults.com/Weather-Extremes/US/hottest-cities.php.

10 of the hottest cities in the US - AccuWeather. https://www.accuweather.com/en/weather-news/10-of-the-hottest-cities-in-the-us/432421.

22 Interesting & Important Facts About The Weather You Should Know.
https://ownyourweather.com/facts-about-the-weather/.

73 Interesting Facts About United States - The Fact File. https://thefactfile.org/united-states-facts/.

The Most Anomalous Weather Events in U.S. History (Part 1).
https://www.wunderground.com/cat6/Most-Unusual-Weather-Events-US-History-Part-1.

This is the US: 56 amazing facts about country, citizens and customs. https://www.usatoday.com/story/money/2020/03/06/56-most-amazing-things-about-america-today/41212433/.

The 21 Strangest Things About 2021's Weather. https://weather.com/news/news/2021-12-08-strangest-weather-2021.

Interesting Facts About the United States - GeeksforGeeks. https://www.geeksforgeeks.org/interesting-facts-about-usa/.

Family

Mom's babies born on 8-8-08, 9-9-09, 10-10-10 – Deseret News. https://www.deseret.com/2010/10/15/20147039/mom-s-babies-born-on-8-8-08-9-9-09-10-10-10/.

The Coble family's tragedy and miracle - CNN.com. https://www.cnn.com/2011/LIVING/01/18/o.coble.tragedy.miracle/index.html.

Families in the United States - Statistics & Facts | Statista. https://www.statista.com/topics/1484/families/.

5 facts about the modern American family | Pew Research Center. https://www.pewresearch.org/short-reads/2014/04/30/5-facts-about-the-modern-american-family/.

66 Fascinating Facts About Family That You Never Knew - Facts Crush. https://www.factscrush.com/2023/03/facts-about-family.html.

How the American Family Has Changed | Pew Research Center. https://www.pewresearch.org/social-trends/2023/09/14/the-modern-american-family/.

How Technology Affects Family Dynamics and Social Development. https://care-clinics.com/how-technology-affects-family-dynamics-and-social-development/.

Military

48 Interesting U.S. Military Facts | FactRetriever.com. https://www.factretriever.com/us-military-facts.

34 Fascinating Facts About the Army That Deserve a Salute - Best Life. https://bestlifeonline.com/crazy-army-facts/.

20 Military Facts That May Surprise you - Westgate Resorts. https://www.westgateresorts.com/blog/21-military-facts/.

33 Army Trivia Facts That May Surprise You. https://www.uso.org/stories/1546-33-military-facts-that-may-surprise-you.

11 US military history facts that might just intrigue you - Sandboxx. https://www.sandboxx.us/blog/11-u-s-military-history-facts-that-might-just-intrigue-you/.

48 Interesting U.S. Military Facts | FactRetriever.com. https://www.factretriever.com/us-military-facts.

33 Army Trivia Facts That May Surprise You - United Service Organizations. https://www.uso.org/stories/1546-33-military-facts-that-may-surprise-you.

20 Interesting Military Facts - Facts.net. https://facts.net/general/20-interesting-military-facts/.

10 Interesting Military Facts For Armed Forces Day. https://www.vehiclesforveterans.org/10-interesting-facts-for-armed-forces-day/.

5 mind-blowing facts about the US military - We Are The Mighty. https://www.wearethemighty.com/articles/mind-blowing-military-facts/.

Quirky Towns and Cities

The Brazilian Town Where the American Confederacy Lives On - VICE. https://www.vice.com/en/article/gq8ae9/welcome-to-americana-brazil-0000580-v22n2.

The 20 Quirkiest Cities in America - Travel + Leisure. https://www.travelandleisure.com/travel-tips/so-youre-a-little-weird-the-20-quirkiest-cities-in-america.

The 20 Quirkiest Towns in America | TIME. https://time.com/3206951/quirkiest-towns/.

The 50 Most Charming Small Towns In America | EnjoyTravel.com. https://www.enjoytravel.com/us/travel-news/places-to-visit/americas-most-charming-small-towns.

The 12 Quirkiest Small Towns in the U.S. — Best Life. https://bestlifeonline.com/quirkiest-small-towns-us-news/.

12 of the Strangest Towns in the U.S. | The Discoverer. https://www.thediscoverer.com/blog/the-strangest-towns-in-the-u-s/Y_js8R6QgAAH77Jc.

The Weirdest Small Towns in the United States - Ranker. https://www.ranker.com/list/weird-american-small-towns/noah-henry.

50 American Small Towns Known for the Weirdest Things. https://www.rd.com/list/american-small-towns-weirdest-things/.

https://bing.com/search?q=odd+towns+in+the+USA.

Orlando. https://www.bing.com/travel/place-information?q=Orlando&SID=8eea2eec-4a57-4f52-b19d-0a2cde213716&form=DCTCAR.

Las Vegas. https://www.bing.com/travel/place-information?q=Las+Vegas&SID=26dfb75a-3573-4ff8-bbb3-b8cadaea23a8&form=DCTCAR.

Conspiracy

The 20 weirdest mysteries that are still unsolved - Shortlist. https://www.shortlist.com/news/weirdest-unsolved-mysteries.

History Decoded: The 10 Greatest Conspiracies of All Time - Brad Meltzer.
https://bradmeltzer.com/TV-Kids-and-More/History-Decoded.

50 Famous Unsolved Mysteries And Spooky Cases - Parade.
https://parade.com/1194770/marynliles/unsolved-mysteries/.

30 Unsolved Mysteries that Fascinate Americans — Best Life.
https://bestlifeonline.com/unsolved-mysteries/.

Takeaways From the AP's Look at the Role of Conspiracy Theories in ....
https://www.usnews.com/news/politics/articles/2024-01-31/takeaways-from-the-aps-look-at-the-role-of-conspiracy-theories-in-american-politics-and-society.

Takeaways from the AP's look at the role of conspiracy theories in ....
https://apnews.com/article/conspiracy-theory-misinformation-trump-qanon-facebook-twitter-39505e8c05b55f91856d809ab6e553a8.

Modern   Conspiracies   in   America: Separating   Fact   from   Fiction.
https://rowman.com/ISBN/9781538164631/Modern-Conspiracies-in-America-Separating-Fact-from-Fiction.

Printed in Great Britain
by Amazon